# Mineral resources and their management

## THEMES IN RESOURCE MANAGEMENT
Edited by Professor Bruce Mitchell, University of Waterloo

**JOHN BLUNDEN**

# Mineral resources
# and their management

**Longman**
London and New York

Longman Group Limited
Longman House, Burnt Mill, Harlow
Essex CM20 2JE, England
Associated companies throughout the world

Published in the United States of America
by Longman Inc., New York

© Longman Group Limited 1985

First published 1985

**British Library Cataloguing in Publication Data**
Blunden, John
   Mineral resources and their management.
   (Themes in resource management)
   1. Mines and mineral resources—Management
   2. Mineral resources conservation
   I. Title  II. Series
   333.8'5    HD9506.A2

   ISBN 0-582-30058-4

**Library of Congress Cataloging in Publication Data**
Blunden, John.
   Mineral resources and their management.

   (Themes in resource management)
   Bibliography: p.
   Includes index.
   1. Mines and mineral resources.  2. Conservation of
natural resources.  I. Title.  II. Series.
TN153.B58  1985     333.8'5     84-10061
ISBN 0-582-30058-4

Set in Times Roman 10/12 on Penta System
Produced by Longman Group (FE) Limited
Printed in Hong Kong

*For Margaret*

# Contents

# List of figures

List of figures

# List of tables

# Acknowledgements

I owe a considerable debt to the many people with whom, over a decade or more, I have discussed aspects of mineral resource exploitation and management that appear in this book. These began in earnest in the early 1970s when I was Research Fellow with the Mining Environmental Research Unit at Imperial College in London and have continued through a participation in some of the valuable symposia organised by the Institution of Mining and Metallurgy.

More recently I am glad to be able to acknowledge my appreciation of discussions with fellow members of the Canadian Land Reclamation Association; with individuals from the United States Bureau of Mines and the Ministry of Natural Resources, Ontario; and those associated with the Centre for Resource Studies at Queen's University, Kingston, Ontario, not least Professor G. K. Rutherford also of the Department of Geography at that University. The Centre continues to produce much that is best in research into minerals policy issues.

But I must also turn to another Ontario university, Waterloo, to express my thanks to its professor of geography, Bruce Mitchell. As series editor his meticulous consideration of my draft text, and the sound and constructive advice he has proffered since, have been of immense help in the final preparation of this book. As on so many previous occasions, I am also happy to record the kind assistance of John Hunt in the execution of the maps and diagrams and Paul Smith in the checking of references to other works cited in this book. However, its realisation in the last analysis owes much to the tolerance, forbearance and support given by my wife and family. To them I am especially grateful.

John Blunden
Oxford
February 1984

# Foreword

The 'Themes in Resource Management' series has several objectives. One is to identify and to examine substantive and enduring resource management and development problems. Attention will range from local to international scales, from developed to developing nations, from the public to the private sector, and from biophysical to political considerations.

A second objective is to assess responses to these management and development problems in a variety of world regions. Several responses are of particular interest but especially *research* and *action programmes*. The former involves the different types of analysis which have been generated by natural resource problems. The series will assess the kinds of problems being defined by investigators, the nature and adequacy of evidence being assembled, the kinds of interpretations and arguments being presented, the contributions to improving theoretical understanding as well as resolving pressing problems, and the areas in which progress and frustration are being experienced. The latter response involves the policies, programmes and projects being conceived and implemented to tackle complex and difficult problems. The series is concerned with reviewing their adequacy and effectiveness.

A third objective is to explore the way in which resource analysis, management and development might be made more complementary to one another. Too often analysts and managers go their separate ways. A good part of the blame for this situation must lie with the analysts who too frequently ignore or neglect the concerns of managers, unduly emphasise method and technique, and exclude explicit consideration of the managerial implications of their research. It is hoped that this series will demonstrate that research and analysis can contribute both to the development of theory and to the resolution of important societal problems.

John Blunden's *Mineral resources and their management* is the fifth book in the series. It addresses many of the issues identified as central to the series through an excellent mix of concepts and information. The fundamental role of minerals in global and regional economies is clearly demonstrated. Research, management and development opportunities are identified while existing and potential problems are described and assessed.

In the first chapter, the discussion of the controversy at Lee Moor in England illustrates the diversity of issues associated with management and development of minerals: understanding the origins of mineral deposits and establishing available reserves relative to demand and production costs; selecting the appropriate extraction process; processing the extracted ore, disposing of waste and rehabilitating the production site; distributing the product; establishing the role of the mineral in the regional or national economy; and determining the value of the mineral resource. The Lee Moor debate emphasises the variety of technological, economic, environmental and political considerations which arise in the management and development of minerals.

Blunden distinguishes between the concepts of mineral resources and reserves. By noting the assumptions that must be made in establishing resource/reserve relationships, he emphasises the uncertainty with which those in the mineral industry must cope. These uncertainties are compounded when estimates are made regarding short and long-term demand for minerals. However, Blunden effectively alerts us to the problems of estimating supply and demand before describing and assessing the future prospects for key minerals ranging from iron, nickel, tungsten, aluminum, copper, tin, gold and uranium to sand and gravel, clay, potash, and salt as well as to oil, natural gas and coal. By carefully assessing the strengths and weaknesses of available evidence and forecasts, Blunden is able to suggest the minerals for which shortages may be expected in the years ahead.

In discussing future prospects for satisfying demands for minerals, Blunden shows how inter-related the world economy is. Many minerals move among and between nations at a global scale. This situation highlights the political and military significance of selected minerals. It also shows that management and development decisions are often made in a much wider context than the minerals per se. As a result, it stresses that understanding of investment and decision patterns associated with minerals often is possible only with an appreciation of geopolitical issues in the world.

A distinctive feature of the book is that after presenting the evidence on supply and demand, it examines the role of minerals in regional economies and the environmental consequences of mineral exploitation. This two-fold focus helps to facilitate an assessment of the benefits and the costs of mineral development. Furthermore, Blunden shows that mining wastes themselves often can become useful products. He does not minimise some of the serious negative externalities created during mineral extraction, but does indicate how in some situations problems can be transformed into assets.

*Mineral resources and their management* is a topic for which John Blunden is well qualified. For years he has studied mineral resources, and in the process has travelled to a large number of countries to examine their problems on a first-hand basis. He also has served as a consultant to the public and the private sectors concerned with minerals. As a result of this wide experience, he brings an informed, perceptive and balanced perspective on the vitally important topic of mineral resources.

Bruce Mitchell
University of Waterloo
Waterloo, Ontario

February 1984

# CHAPTER 1

# Introduction and overview

## 1.1 The Lee Moor controversy

The following pages present a sequential collection of accounts taken from the *Western Morning News*, a daily paper which circulates in the south-west of England, relating the progress of a public inquiry by the Department of the Environment which took place in the early 1970s. Its purpose was to take evidence from all interested parties as to whether English China Clays Lovering Pochin and Co. Ltd should be allowed to work china clay (a mineral primarily used in the manufacture of fine china and in the production of paper) and to tip the very considerable quantities of waste resulting from its extraction in 237 hectares of Dartmoor. The area is not only adjacent to workings and processing plant already in use for china clay production but, of more significance in terms of the controversial nature of the proposal, would encroach on 123 hectares of land inside the south-western boundary of Dartmoor National Park (Figs 1.1 and 1.2). The task of the Department of the Environment inspector responsible for conducting the inquiry was subsequently to interpret the evidence with a view to recommending to the Secretary of State for the Environment a particular course of action regarding the land in question.

In the UK, changes in land use from one purpose to another have, since 1947, required the consent of the relevant local planning authority. Where the applicant and the planning authority cannot agree to a course of action or where the issue of land use change is particularly contentious, the planning authority will usually refer the matter to the Secretary of State for the Environment for a decision. His normal recourse before committing himself will be to call for a public inquiry simply because such an undertaking should allow the maximum opportunity for the expression of all shades of opinion regarding the proposed land use change.

How such an inquiry works out, in fact, is well enough illustrated by the narrative which follows, but the primary purpose of these daily accounts of

1

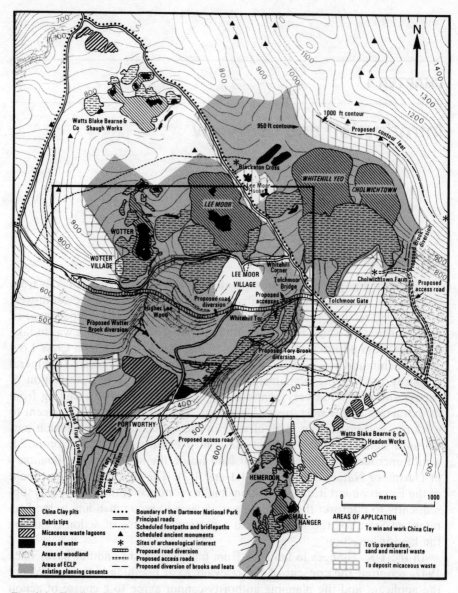

*Fig. 1.1* English China Clays Lee Moor Area Planning Application. *Source*: Blunden (1977)

the Lee Moor controversy and its associated public inquiry is to draw attention at this early stage in the most vivid way possible to most of the central issues that will be later examined in greater detail in this book.

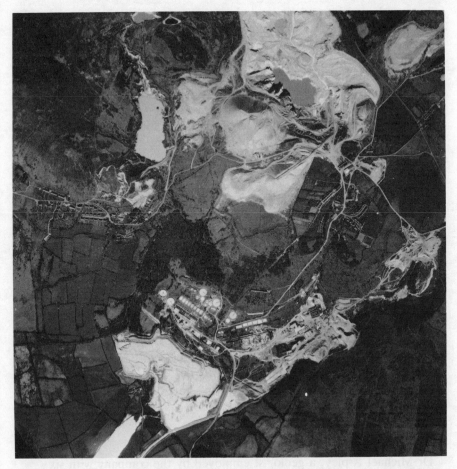

*Fig. 1.2* An aerial photograph of the Lee Moor china clay working immediately before the public enquiry concerning its extension. *Source*: Aerofilms

## 1.2 Progress of a public inquiry: resource file on the Lee Moor controversy

*Western Morning News*

### China clay landscape 'to be reshaped'

Over the next 50 years the china clay industry in the Lee Moor area will, if the present application by English Clays Lovering Pochin & Co. Ltd, is allowed, produce 26,365,200 tonnes of clay and 201,625,200 of waste products.

This massive volume of waste consisting of micaceous residue, overburden and quartz sand will at the end of the 50 year period have

3

been reshaped into a man-made landscape of merit in its own right, the Lee Moor China Clay public inquiry was told at Exeter yesterday.

One of the company's expert witnesses, Mr Gordon Ley, a mining engineer concerned with development planning, said the sporadic development and uncoordinated waste tipping of past years would be brought into line under the plan for future development. Ultimately there would be two major open pit excavations – one in each area of claybearing ground – and integrated tipping in four areas. All other tips would be removed during the course of working.

## Disused area

Mr Ley said micaceous residue disposal would extend the existing mica dams immediately to the north of Portworthy dam. In this way, he said, the number of existing isolated tips and dams could be reduced and the much larger tip so formed could be adequately phased to try to reduce their impact on the environment as much as possible, by progressive shaping and planting. A new method of conveyor belt tipping, forming sand tips in a series of 24.4 metre vertical lifts, the toe of each being inset 9.14 metres from the lip of the layer it overlays, would give a more stable tip and would have environmental advantages, said Mr Ley.

Vegetation of the lower levels of the tips could be started while the higher levels were being tipped, and the overall slope of the tip would be reduced from 30 to 22 degrees to the horizontal, which more approached the natural slope of the Moor.

## Rare commodity

Dr Michael Ripley, a geologist employed by the company with six years' experience in the china clay industry, said china clay deposits of high quality and in sufficient quantity for commercial development existed only in Georgia, United States, and the south-west of England. High quality china clay was a rare commodity on a world scale, he said. Extended workings from the 290 to the 305 metre contour would release the clay between those contours and release the clay lying beneath the 290 metre contour. This would allow the working of the pits to a greater depth.

Working to the 305 metre contour would release 1,727,200 tonnes of clay, most of which was the highest quality china clay in the Dartmoor area.

## Situation at clay pit critical

Crucial increases in china clay production in the Lee Moor area are dependent upon new planning permissions for land on which to tip the millions of tonnes of waste sand, rock and other material which is a by-product of the industry.

Mr Gordon Ley, mining engineer with English Clays Lovering Pochin & Co. Ltd said that the situation was most critical at the Whitehill Yeo pit, where there was no capacity available for tipping sand or overburden.

Similarly, there was no capacity for Cholwichtown overburden and the life of Cholwichtown sand tip was only 13 months. The life of the Lee Moor overburden tip, together with the Blackalder Tor tip, was about $13\frac{1}{2}$ years and geared to the production of the Lee Moor pit alone, which was at present time less than one-third of the production of the whole complex.

If additional possible areas were brought into use, two and a half years of tipping from Cholwichtown and Whitehill Yeo pits would be possible.

Mr Ley said that assuming complete flexibility, given that either overburden or sand from any pit could be tipped on any tip, total tipping life in the complex is five years. Some 203,200 tonnes of sand were sold annually, but there was no potential or existing market for the fine micaceous residue, of which 381,000 tonnes was produced annually. The company, therefore, allowed in its proposals for the whole of the anticipated production of this residue – 19.35 million tonnes over 50 years – to be retained in mica dams or used for backfilling disused areas of the workings. One of these, the worked-out Wotter pit, would eventually be filled with residue and become a shallow lake. The effects of the concurrent vegetation of the new-style 'layered' tips and other landscaping should be evident when the first stage of the company's three-phase plan is completed in 1984.

## 50-year china clay forecasts queried

A mining engineer of English Clays Lovering Pochin & Co. Ltd yesterday denied that there was any 'air of unreality' about the company's 50-year plan for the Lee Moor china clay complex. The engineer, Mr Gordon Ley, was cross-examined on his company's proposals by Mr W. Evans, of Plymouth City Council, during the Lee Moor china clay public inquiry at Exeter.

'Given production figures for five years', said Mr. Evans, 'you work out a plan for 30 years. Is not there an air of unreality about these proposals?' Mr Ley replied that on the basis of production forecasts they had used, the plan they had worked out would last so long. 'The important thing is one does not change the relationship between the after treatment and the mining in changing the production forecast', said Mr Ley. Mr Evans said it is conceivable that it may be economic

5

to abandon certain parts of the workings, and there would be left a half-completed landscaping plan.

## Cross examined

'This is not conceivable', said Mr Ley. 'We must get this in proportion. In terms of china clay deposits the whole of Lee Moor is an area of very high quality clay reserve.' Mr Ley was then cross-examined by Lady Sayer of the Dartmoor Preservation Association, about the sale of sand from the Moor over the next 50 years. She asked whether the company could not sell more than 10.16 million tonnes during that period. He said there was no foreseeable increase in the market for sand, and therefore, the company's proposals were designed to meet the worst possible position. There was nothing to stop the proposals being modified downward if a large quantity of sand could be disposed of. Lady Sayer said one was constantly reading of the acute shortage of sand and gravel in the southern part of England and London. Surely it was possible the company's figures might well be dictated eventually by shortages in other areas. 'Maybe', said Mr Ley.

## Dumping ground

'Your company's plans are really directed to using a large area of southern Dartmoor for a waste dumping ground?' she asked. 'Yes, if you wish to express it in those terms', replied Mr Ley. Lady Sayer then referred to the Wotter pit, which will be worked out by the end of the first 12 year phase of the plan, then become a lake, and finally a mica dam with a few feet of water to preserve the appearance of a lake.

## Death trap

'This slurry covered with water could surely be a "death trap" to paddling children' said Lady Sayer, and would have to be fenced. Mr Ley said there was no attempt to suggest a veneer of water would be left to tempt children to paddle. There was no definite proposal for the water. It could be two or three feet deep, or deeper if necessary.

## 'Dalton's Moor' fears at inquiry

Dalton's Dartmoor not Nature's Dartmoor would be the result of English Clays Lovering Pochin's restoration plan for the Lee Moor china clay complex, the public inquiry at Exeter was told yesterday by Lady Sayer, representing the Dartmoor Preservation Association and other amenity bodies.

Mr Alan Dalton is managing director of E.C.L.P., whose proposals to work and tip on another 242 hectares of land at Lee Moor, 123.43

hectares of which are in the National Park, are the subject of the inquiry. Lady Sayer was cross-examining Professor A. D. Bradshaw, a specialist in vegetating waste material, called in by the company.

### Good grazing land

He said there was no reason, given proper fertilisation, why the tips and mica dams should not be able to support grass and trees. Some grass areas could become reasonable, even good agricultural grazing land capable of being farmed profitably. There was a possibility of waste areas being covered with growth within six months of being released from working. Land reclamation could not be carried out in one fell swoop. Five years of after care were needed, with reviews up to ten years.

The landscape surveyor's department at E.C.L.P., under Mr L. D. Owen, was essential because someone had to be on the spot to care for what was being done. The professor pointed out that china clay workings had a no more catastrophic effect than the original clearing of forest cover from Dartmoor and establishment of hedge banks. Time had healed these disturbances and now people were happy with what they saw.

The new tips would be dealt with as soon as they were made. A vegetation cover and landscape could be brought in immediately, with the first trees and grass being planted on the tip perimeter which would be established at the outset. Mr. Michael Ripley, a geologists with E.C.L.P., was cross-examined by Lady Sayer about the future of the Lee Moor workings after the end of the fifty year period. She asked whether, if more clay was found, there would be more applications for further tipping areas. Mr Ripley said the company would not use the clay, if found, until the end of the 50-year period because their market forecasts were already taken care of. Lady Sayer asked whether this was, in fact, not the end of things, and they may find another application coming in another 50 years time. 'It could be', said Mr Ripley. 'It depends on how much clay we find.'

### Jobs warning

The result of the Lee Moor inquiry is of vital interest to the 1,000 people involved in the china clay industry in the area. These people earn between them about £1.5 million a year from the industry – and the village of Lee Moor is dependent upon it. A warning was given at the inquiry on what could happen if the decision went against the company. 'Men now employed in the Devon area would find it difficult to find alternative work, if, as the result of the decision on the planning application, we were forced to cease operations', said area manager Mr William Stevens. 'Most are family men, with dependents enjoying earnings which are relatively good for this part of the country', he said. 'If we do not get permission for more tipping area for the near future we would in some cases have to tip waste on our clay reserves, and we would eventually be shut down prematurely.'

## Lee Moor area 'A country park'

Preservationists anxious about the mounting pressure on Dartmoor from the motoring public could, although a long time ahead, have an ally in the Lee Moor clay workings. As consultant landscape architect Mrs Sheila Haywood sees it, the whole English Clays Lovering Pochin industrial complex would, after the proposed 50-year development plan, have undergone a complete change. Her 'ultimate objective' envisages the area as a country park which would take the public pressure off Dartmoor. She told the china clay inquiry at Exeter yesterday that the three-phase plan showed how the existing and proposed workings added up to a major change in the area's landscape. 'Although this application is mainly concerned with additional facilities for tipping, the company is anxious to demonstrate that the changes, not only from this application, but the whole of the existing workings for which there is no commitment for overall landscaping, can be controlled and moulded towards an acceptable "after use" and an acceptable future landscape.'

### Exciting

'I feel personally that the future landscape could be an extremely stimulating and exciting one which would in no way detract from the character of Dartmoor itself. In two respects it could be of benefit to Dartmoor. Firstly, the lake areas could be used as reservoirs, thus relieving pressure on possible sites within the National Park. Secondly, the whole after-use area could be considered as a future country park, catering for more intensive public use than is desirable in the National Park itself. Lee Moor, being five miles from Plymouth on that side of the Moor, is well situated to draw off public pressure on Dartmoor', said Mrs Haywood. The first phase of the plan – which would end after 12 years – is the one which she thinks represents the period of greatest upheaval. 'This is the phase within which the outlines for the future are laid down. The next two phases, although a great deal goes on within them, represent to a rather greater extent the stages of consolidation along lines which have already begun to emerge in phase one.'

### Three lakes

The second phase of the plan anticipates development up to 30 years hence and the third phase up to 50 years. Three lakes would be formed at the end of the 50 year period. The Lee Moor and Shaugh Moor

Lake would be 3.62 kms long, the Hemerdon Headon pits lake would be 1.6 kms long and the Wotter Lake would be just over 0.4 km long. During the three phases a total of 460 hectares of the complex would be rehabilitated with grass, trees and shrubs, said Mrs Haywood.

## Lee Moor scheme 'for future generations'

The Lee Moor china clay public inquiry is into a scheme of which very few people now attending the hearing will see the final result. The development plans 'the ultimate landscape objective' of the English Clays Lovering Pochin industrial site as a future country park, but that will not be until 2031 or 2036. The time factor was mentioned at yesterday's hearing at Exeter by Mr Douglas Frank, QC, representing Devon County Council, who said he was concerned that future generations were being legislated for. Lady Sayer, representing the Dartmoor Preservation Association and two other amenity bodies, also expressed concern about the possibility of the scheme stretching over a greater period of years than envisaged. She said it had been stressed many times that the three phases of the 50-year plan were linked to production and sales and this could mean, if demand and sales should undergo a recession, that the phases might continue for far longer than 50 years.

### In 100 years

As an extreme example, she said there would still be disturbance of the Dartmoor landscape over a wide area in 100 years' time. Mrs Sheila Haywood, the company's consultant landscape architect, who was being cross-examined about her proposals for the area, said the time scale to which the restoration plan was geared was the best judgement the company had been able to make of the situation. The whole problem had been examined on a big scale, and the 50 years had emerged as a realistic basis for the plan. Mrs Haywood said there would clearly be a great deal of change if the company's proposals were allowed, and her aim was to make them as acceptable as possible.

One of the changes, about which Lady Sayer had expressed concern, was the cutting of a new road through Higher Lee Wood. Mrs Haywood said this would provide a great opportunity for more people to enjoy it. It was one of the brighter spots of the scheme. Asked if she could give a guarantee about the preservation of the ancient monuments the company had undertaken to preserve, Mrs Haywood said the most effective means of protecting them would be to draw them as much as possible to the attention of the management and so far it was possible the company would care for them.

9

*By the back door*

The creation of a lake 3.62 kms long from the eventual filling of the
Lee Moor and Shaugh Moor pits as an ultimate landscape objective
was really 'constructing a Swincombe reservoir' by the back door,
claimed Lady Sayer. The lake which would be created would be bigger
than the Swincombe reservoir rejected by Parliament.* The company
had suggested it should be used for reservoir purposes, relieving
pressure on possible sites in the National Park. 'What is really being
proposed is something not for the good of Dartmoor, but in order to
mitigate a huge tipping scheme', said Lady Sayer. Mrs Haywood
replied that the company had tried to reconcile opposing interests. The
Lee Moor area could eventually be a help to Dartmoor acting as a
'buffer' between Plymouth and the Moor itself. It would absorb a large
number of people who wanted a day in the country. 'This could go on
forever', said Lady Sayer. 'You could take another step into the
National Park and call that a buffer between Lee Moor and the rest of
Dartmoor. In the end, you would be left with nothing.'

**Clay chief says 'we hope to conserve'**

Although the preservationists would probably not agree, Mr Alan
Dalton, managing director of English Clays Lovering Pochin & Co.
Ltd, says that his company's proposals for the Lee Moor area are an
attempt to interpret practically the definition that conservation is 'a
wise use of nature and of natural resources'. He told the Lee Moor
inquiry at Exeter yesterday that his company was conscious that there
was a clash of interest over Dartmoor and the minerals lying under it.
But both the minerals and the open spaces were needed, and the
development at Lee Moor was a coupling of modern methods of
working with a sympathetic landscaping treatment, which would
produce an ever-diminishing scale of exposure. The ultimate landscape
in his opinion, would be neither unsightly nor incongruous and would
in due course be returned to the nation's stock of open park-land. The
company's marketing director, Mr Thomas Pleasants, said in order to
keep the fast growing European paper industry supplied from the
United Kingdom, rapid production increases have been necessary.

*Existing pits*

The process of enlarging existing pits and bringing new ones into
operation will have to continue if this country is to retain its share of

---

*A proposed major water storage project which would have been located in the
Dartmoor National Park, but is now defunct.

---

the market. The 14 per cent of the company's production from Devon pits would have to be increased in view of the 50 per cent growth in total demands for china clay by 1975. Devon clay was nearer the home market, and therefore cheaper within this country.

Mr Pleasants warned that if there was any restriction upon the production growth of the Lee Moor operation the effects on both customer and the industry would be profound and irrevocable.

(*Source*: *Western Morning News*, 28 September – 14 October 1971)

## 1.3 Identifying key issues

Of fundamental concern in the above narrative must be the *origins* of the deposit of china clay and how it came to be located where it is. Like all the minerals that are exploited, they have taken tens to hundreds of millions of years to form in the earth's crust. Sometimes the remains of formerly living organisms have resulted in the creation of minerals. The fossil skeletons of small invertebrates have given rise to limestone. The preservation of carbohydrates from plants has led to the formation of coal, and hydrocarbons from marine organisms has led to the formation of oil. In other cases, such as salt, the original material was held in solution in sea-water and evaporated out by the action of sun as seas became shallower.

Most minerals, like all other rock formations, originated from igneous rocks, the material from which the first solid crust of the earth was formed. Some minerals are found in the igneous material itself, some in the sedimentary materials formed from these through the action of weathering and erosion, and some in the metamorphic rocks which may be igneous in origin or sedimentary deposits that have been transformed by heat or pressure or both. Where the minerals are concentrated as ore bodies they were formed by the 'settling out' of the minerals as the molten igneous material (magma) cooled beneath the earth's surface. Both solutions and gases from cooling domes of magma have often penetrated overlying sedimentary materials along lines of weakness, creating in some instances what is termed a metamorphic auriole. These are frequently characterised by consecutive bands of different minerals reflecting the varying temperatures at which the gases and liquid condensed and solidified. Around Dartmoor a typical example of such an auriole may be seen with tin and wolfram (tungsten) closest to the now exposed granitic dome and arsenic and copper further out, whilst china clay at Lee Moor is the result of the magmatic gases breaking down the granitic materials. The exact process is described in greater detail at the beginning of Chapter 2, 'Defining Resources'. Where all the other minerals are concerned, their geological origins are described in a much less generalized way in Chapters 5, 6 or 7.

A second issue is the *magnitude of the workings* in question that will yield nearly 26 million tonnes of clay and almost 200 million tonnes of waste and the way in which the material is won, that is the *mining techniques* used. The extraction of china clay is by open-pit excavation (one of the mineral extraction techniques most commonly in use). This method of extraction is particularly well suited to any unconsolidated type of deposit, whether it be china clay, or sand and gravel, or a disseminated metallic ore, so long as the depths involved are not great. Open-pit extraction techniques offer the advantage of operational flexibility, are safer than the other major form of minerals extraction, underground working, and are economically more attractive since they permit the application of earth moving machines at a scale which allows for maximum cost effectiveness.

Before operations such as those at Lee Moor can begin they require the removal of the overlying unwanted rock (or overburden). When this is added to other waste materials (in the case of china clay, quartz and micaceous residues), the amount of waste extracted compared with valuable mineral can rise to around 88 per cent, though more typical figures for open-pit ore mines would be 40 per cent and for strip mines around 80 per cent (Dubnie 1972; Blunden 1975). The difference between strip mines (mainly used to extract coal) and other forms of open-pit operation is that the former involves the *progressive* removal of overburden, prior to the extraction of the mineral and the return of the overburden to the site. Although this overburden replacement did not happen in the past, it is now a common requirement that even the surface soils are put back in place, and the drainage and vegetational cover restored. Indeed, given the emphasis at the Lee Moor public inquiry on *site rehabilitation* (since restoration is not in this instance practicable), it is not surprising that the need for some satisfactory arrangements regarding the landscaping of a mining area once the mineral extraction ends is now a commonplace requisite in developed countries if consent is to be given for the initial working. Moreover, the idea of having a public inquiry at all must be indicative of a *wider framework of control inside which mineral operations take place*, a subject addressed in Chapter 8.

The other most important extraction technique, apart from the open-pit method, is that of underground working. Whilst this could not be used to extract china clay because of the nature of the material and its location in deep granitic pockets close to the land surface, underground methods fall into two distinctive types. Where 'open stoping' is used, the rock body from which the valuable material is extracted is sufficiently strong to permit underground cavities to remain after the ore has been removed through drilling and blasting. The other approach, that of 'filled stoping', means that all waste materials including, in the case of metalliferous minerals, the tailings (that is the fine waste rock left after the valuable ore has been removed) are used to fill those portions of the mine that have been worked out. This provides sufficient ground support to allow further extraction of the ore. The back filled material is usually mixed with cement to improve its strength (Thomas 1973).

Another means of minerals extraction involves the drilling of bore holes from the surface to permit the removal of the valuable material. In the case of oil and natural gas, where these are trapped in a zone of sedimentary rocks, they usually emerge of their own volition under pressure, though for the winning of salt, potash and sulphur, the injection of water results in the creation of a solution which can then be pumped to the surface. Where oil is extracted from tar sands, steam is used to free the mineral from the parent rock body, before it is pumped to the surface.

All these methods of extraction are related to the winning of the various minerals discussed in Chapters 5, 6 and 7. However, it has to be emphasised that in terms of the amount of wastes produced from underground workings, as compared with their open pit counterparts, the former gives rise to far less surface disturbance from their disposal. For example, the quantities of waste produced by sub-surface iron mining are very small and even in the case of non-ferrous metals, these amount only to between 10 and 20 per cent of the total material worked. Insofar as the metallic ores are concerned the reason for this situation lies in the fact that in the majority of underground mines the workings follow specific veins of concentrated metal ore. In contrast, the waste produced by surface mining can easily be 50 times greater than that of underground workings with those for china clay at the top end of the range.

As emerged in the account of the public inquiry, much is made of the question of wastes and what is to be done with the quartz tips (ultimately made up from around 182 million tonnes of material) and the micaceous residues penned back by dams in the fashion of tailings (over 19 million tonnes). Such waste materials with the exception of overburden are almost entirely derived from the *processing* of the extracted material to remove the valuable minerals.

China clay is perhaps amongst the simplest of all minerals to process since unrefined kaolin is extracted by water jet from the working face of the pit in the form of a slurry with the quartz fragments and micaceous residues settled out from the valuable mineral in nearby tanks. Only sand and gravel which involves on-site washing and perhaps some crushing once it is dug from the pit, is more straightforward. This contrasts with metallic ores. There, the beneficiation process in which the unwanted constituents of the ore body are removed is again carried out on site because of the relatively low value of the untreated ore. Only in the case of the most valuable minerals (or those mined in high concentrations such as iron), would transport over long distances be envisaged for processing at the market end.

At the beginning of what is a three-stage process, the ore is comminuted by crushing and/or grinding. It is then concentrated, a means by which the desired ore is separated from other materials (commonly called gangue) which are then discarded as waste. Separation may be carried out by one of three methods. Gravity separation, the means used for china clay, is deployed to concentrate some metallic ores (such as gold and silver) as well as coal, but for other minerals it has been largely replaced by flotation. This method of concentrating mineral ores is based on the principles of surface chemistry and is most widely used with metal sulphide ores, though it competes with other separation processes in the beneficiation of a wide range of other metallic ores, coal and industrial minerals. The chemicals which are used to 'float out' the ore from the fine rock particles are currently selected not only for their effectiveness but for those qualities which may do least harm to the environment should they be discharged into local water courses; that is, they have low toxicity, rapid degradability, lack nutrient properties and have a low content of water soluble toxic metallic salts (Hawley 1972; Blunden *et al.* 1973). However, whilst most flotation circuits are able to recycle their process liquids, all of them contain a settling pond stage which allows solid particles, including metals, to be settled out as well as permitting the treatment of any acids present (Bragg 1975).

Two other approaches to the extraction of metals from the fine rock material employ either magnetic separation (especially for iron ores), and, where the ore is dry, electrostatic techniques (again for iron ores as well as titanium). These methods of processing are mentioned in Chapters 5 and 6 as appropriate. In the case of the metallic ores, by far the majority of waste rock, whether the mineral is won by open-pit or underground techniques, finishes in the tailings pond being the residues of the beneficiation processes. Whilst metallic ores have become leaner over recent years and the amount of tailings to be disposed of has grown, it is still true to say that the lower the grade of ore that can be worked economically, the higher the waste content. Thus bituminous coal operations produce about 35 per cent waste, most of which is in the form of tailings, potash 51 per cent, asbestos 97 per cent and uranium 99 per cent. The comminuted fines resulting from processing ore are less dense and occupy more volume than the parent rock. The average size of tailings area per tonne of material extracted runs out at around 210 square metres for Canadian mining operations, a calculation based on a consideration of a wide range of non-ferrous metalliferous workings but not china clay.

Whilst china clay tailings are inert, those resulting from the beneficiation of metal ores are impregnated with residual heavy metals and sulphur and may even contain radio active materials. This can give rise to a number of problems, particularly when the tailings site reaches the end of its useful life. At such a stage, unless properly controlled, the water run off from such areas (often in the form of a weak sulphuric acid) can severely damage ecosystems if it enters the drainage system. Moreover, wind-blow from the surface of such areas can prove an environmental hazard where efforts have not been made to control the surface, usually by attempting to establish some form of vegetation cover. These issues of *waste and waste disposal and their interface with the environment* are dealt with at length in Chapter 9.

The further processing of metallic ores can also be a problem of some environmental significance as Chapter 8 demonstrates. Pyrometallurgical techniques for the production of the metal from metallic ores are the most significant contributor here since both the preliminary roasting and the smelting processes frequently drive off sulphur dioxide ($SO_2$). Although the release of the $SO_2$ can be controlled with much of the gas being converted to sulphuric acid, the cost factor and the problems of disposing of the acid means that atmospheric dispersal is still employed.

Two other means of processing metallic ores are sometimes used. Those involving hydrometallurgical techniques extract the metal from the concentrated ore using chemicals in solution, whilst electrometallurgy involves either the heating of the ore in an electric arc furnace or the process of electrolysis usually reserved for the refining of copper, lead or zinc.

Whilst the further processing of metallic ores may thus impact on the environment, those concerned with china clay do not since they merely involve the blending of refined material to suit particular market uses. Indeed, it is really the problem of physical change, the reshaping of the landscape, that the inquiry returns to again and again. Although landscape change is not new (Professor Bradshaw in his evidence suggests that Dartmoor has in the past undergone forest clearance and building of hedge banks to divide the area into suitable units for agricultural purposes), it is perhaps *the scale and speed of change* at a mineral extraction site that causes most concern to those who in developed countries would oppose such operations. This has been particularly apparent since the end of the Second World War as mineral operators have increasingly favoured open pit mines, a factor frequently alluded to in later chapters.

This situation (shown in Table 1.1) has been determined not only by the need to work leaner disseminated metallic ores, but also has been encouraged by the availability of extractive techniques which can lower unit production costs if output is large enough. A similar state of affairs pertains in the production of aggregates and other quarried stone materials. Although world figures concerning quarry numbers are not available because of the

*Table 1.1* World size distribution of mines: 1969/79

| Millions of tonnes of output | 1969 | | 1979 | |
|---|---|---|---|---|
| | Underground | Open pit | Underground | Open pit |
| >3 | 25 | 110 | 50 | 140 |
| 1–2.99 | 145 | 110 | 140 | 140 |
| 0.5–0.99 | 120 | 80 | 115 | 60 |
| 0.3–0.49 | 125 | 70 | 120 | 55 |
| 0.15–0.29 | 165 | 50 | 160 | 50 |
| Total numbers | 580 | 420 | 585 | 445 |

*Note*: The main trend identifiable is not merely the change in numbers of mines but the move to greater size particularly > 3 million tonnes.

*Source*: *Mining Journal Annual Review* (1981).

ubiquity and low value of the materials produced, figures for the UK are indicative of a situation that is likely to relate to most developed countries. Quarry output had risen from over 40 million tonnes in the early 1920s to over 305 million tonnes by 1973 though the number of quarries fell from 6,000 to under 4,000 in the same period (Blunden 1975). The change was particularly apparent from 1948 onwards when the government's Advisory Committee on Sand and Gravel had specifically advocated that local authorities should permit extensions to existing workings wherever possible so as to economise on existing capital investment and to reduce the number of points at which the environment might be affected by such operations.

The intervention of such a committee viewed within the context of a *legislative framework aimed at environmental and land use control of mining activities* typifies the increasing amount of control exercised by governments in developed countries. Indeed the public inquiry system of which Lee Moor is an example is yet another facet of this same phenomenon. That such control impinges on the normal economic relationship between supply and demand in a world market for minerals is evident enough even from the words of ECLP's marketing director who referred at the inquiry to the problems of constraint on his company's activities at Lee Moor in relation to the need for it to maintain its share of a large expanding market at home and abroad which could only be served from two production areas in the world, Georgia (USA) and the south-west of England. Though he was concerned with the constraint of activity made possible by the legislative framework, rather than its promotion, the whole question of what impinges on the free market for minerals is discussed at length in Chapter 4.

As to other issues which were raised at the inquiry, it will be noted that Alan Dalton, the managing director of ECLP referred in his evidence at the public inquiry to china clay as a resource (though he did not define the term). Just what is meant by a *mineral resource and how we can distinguish this from other forms of resource* (since Dartmoor itself might be described as a recreational resource) and why reference is sometimes made to the *reserves of a mineral resource*, are the subject of the two chapters which follow this introduction and cover the defining and classification of resources.

Implicit in many of the exchanges at the public inquiry, however, was the idea that china clay is a scarce resource and whatever the more immediate problems of maintaining adequate supplies for customers, its supply is limited not only at Lee Moor (where an end date for production can be foreseen if certain basic assumptions of a short term nature regarding demand are extrapolated over a longer period), but also at other known production locations. This must suggest a similar situation for all other resources which are consumable and cannot be replaced except over geological time. This raises the question dealt with in Chapter 11 as to *how long our stock of minerals is to last and whether or not possibilities exist to overcome the problems of severe shortages or even ultimate exhaustion.* However, before this in the chapters on specific minerals (5, 6 and 7), the more immediate impact of *spatial variability in the distribution of minerals*, a matter which is explicitly referred to with respect to china clay in the evidence to the public inquiry given by Dr Michael Ripley, ECLP's geologist, is discussed.

Finally, the part played by the extraction of china clay in terms of economic development may be considered an issue of significance. Its importance was hinted at by Dr Ripley, at national level, whilst William Stevens, the manager of the Lee Moor working, spelled out clearly the importance of the extractive operation for local jobs and the injection of cash into the local economy. Attempting to deal with the question of *the effect of mineral exploitation on the economic development of regions* is a matter of major concern in Chapter 10. After a discussion at a global scale it is handled with reference to two major case studies, a non-ferrous metal (tungsten) and a hydrocarbon (oil).

Since this chapter began by using an account of a public inquiry as a means by which to raise some of the key issues to be discussed in this book, it is as well to end it by recording the outcome of that inquiry. When the decision of the Secretary of State for the Environment was made known, the application by the company to work the new area was largely successful though permission was refused for waste-tipping in the major part of the area between Ridding Down and Crownhill Down (Fig. 1.1). Conditions were, however, attached to the consent and these aimed at ameliorating the visual impact of the new operations. Specific schemes of progressive landscaping and site after-treatment when working is completed were required of the company and left to be agreed with the local planning authority.

# CHAPTER 2

# Defining mineral resources and reserves

## 2.1 Resources – A human appraisal

About 290 million years ago, vast earthquake disturbance rammed the Devonian and Carboniferous sedimentary deposits of what was to become Devon and Cornwall against the older rock structures of Wales, crushing and cracking the rocks and causing a massive upwelling of magma. As the magma cooled, vapours emerging from this material, trapped beneath the overlying beds, altered the nature of the felspar, one of its constituents, forming a hydrated silicate of alumina. The sedimentary deposits were eroded over time exposing the granites of Dartmoor, Bodmin and Hensbarrow and the silicate of alumina known as kaolin.

For most of the history of man this material remained where it lay, untouched and seemingly useless. However, in 1745 William Cookworthy discovered that, when fired, kaolin could be used to produce high quality porcelain. Apart from some local usage by Cookworthy, the raw material, now known as china clay, was first sent to the English potteries. As its value for other purposes became known (especially as a filler for paper and as a glaze producing the glossy finish on high quality paper), extraction at the main centres (Hensbarrow and Lee Moor) expanded and an export market developed which by 1980 accounted for 83 per cent of the 2.838 million tonnes produced in Britain.

This story illustrates how a mineral seemingly of no importance can in a relatively brief period become a resource of considerable commercial value. China clay is used as the example here since it refers back to Chapter 1, but a number of other minerals such as uranium, aluminium or titanium could have been chosen. All of these were, at some time, of no intrinsic value until technological advance transferred them to the status of resources. Uranium for example, only recently came to be used as a source of nuclear energy. In the case of the last two minerals, these have only come to the fore in the last forty years as major metallic ores though they had been used earlier as

18

a white pigment in paints. Thus the term resource does not relate to a material or an object. It applies to a *value* placed upon that material or object because of the *function* which that material or object may perform, or the operation in which it may take part.

To become a resource a material or an object must be subject to human appraisal and to be judged as having a value in one role or another. A resource could be defined as such by a single individual but more commonly a general consensus exists regarding the translation of a material or object to the status of a resource, at least so far as those with a shared cultural experience are concerned. Thus whilst members of an industrialised world appreciate the essential value of metallic minerals to their mode of living, this would not be true of other more primitive communities. The Bushmen of the Kalahari, for example, consider moisture in an essentially arid world one of their most valued resources. Moreover they consume a range of plants, grubs and insects, particularly those which are succulent, most of which are quite unpalatable to people in developed societies such as those of the western world. Thus they survive in a world of resources which would not be recognised as such by economically advanced nations and on a resource base far below that which would meet the basic needs of such countries.

From what has been said so far it must be evident that a resource is not static or fixed over time or space – it is indeed as dynamic as society itself. A material once considered a resource of great value can cease to be so. For example, during the Neolithic culture in Britain a site near Thetford in Norfolk, now called Grimes Graves, was for around 1100 years (from 2500 to 1400 BC) a major centre for the mining and working of flints. The society of the period (before metals were discovered) had found that flint could be split and chipped to form sharp edges and was by far the best material from which to make cutting tools (axe heads, knives and scrapers) and weapons (arrowheads). These were dug using antlers and stone hammers from narrow bands of nodules located in the chalk beds via shallow shafts from the bottom of which narrow galleries radiated outwards. Similar sites existed such as those in the Lake District or at Penmaenmawr in North Wales. But as technological development overtook the flint and as copper, bronze and iron (all much more efficient substitutes) subsequently came into use, these stone materials ceased to be of value and within the context of their use as tools and weapons, ceased to be a resource. From being a large centre of industry pocked with shafts, Grimes Graves reverted to an uninhabited area of heathland and birch forest.

Indeed, the history of man in his environment provides a constantly changing picture of resource appraisal. The spread of the Neolithic culture from Western Asia, for example, not only represented one of the most notable events in the history of man's early impact on the environment, but also produced the first major shift in the resource base of man with the emergence of soil as the key factor in the economy. In Britain, Bronze Age man had begun to distinguish between different types of soil, appreciating it not only as a resource in its own right but one of varying quality. It is certain that during the first major clearance of forests, soils deriving from stands of indigenous lime trees were looked upon as a particularly valuable resource.

The reliance in Britain on agricultural resources as the mainstay of economic life continued largely unchallenged until the coming of the Industrial Revolution when technological advance began to make new energy resources available on a massive scale for the first time. These in turn provided the base from which to exploit other forms of material which became recognised as resources. Although the early stages of the Industrial Revolution were supported by the use of water power, the development of pumping and winding gear made the deeper mining of coal possible, thus providing the motive force for the rapid growth of industrialisation. Further technological developments, enabling oil, hydro-electric and nuclear power to become new energy resources, have permitted the development of the highly organised and sophisticated industrial societies of the second half of the twentieth century.

Without large energy inputs the working of low grade disseminated mineral ores by open pit methods would not have been feasible. It is estimated by Chapman (1974) that a body of copper ore of some 100 million tonnes, containing 0.7 per cent metal when worked at a rate of 8,330,000 tonnes per year, is likely to require for mining, crushing, grinding and separation (flotation) a total power input of 40,000 kilowatts per annum. Without such power availability, and the related technological expertise, the low grade copper ore could not have been worked.

This discussion of the nature of a resource requires a distinction to be made between a mineral resource and a reserve. Because power and technology are available to work the resource which is, for example, low grade copper ore, it may be spoken of as part of the total reserves of that resource; in other words, that part of the copper resource that can actually be gained from the earth using existing methods. However, there is a further aspect to the definition of a reserve – the overall price in world markets. Thus whilst copper could be mined underground in plentiful quantities and in veins concentrated at between 5 and 2.5 per cent (as it commonly was around 1900) there would be no incentive to work the much lower grades of ore common today (about 0.4%), since they would be uncompetitively high in price. Greater demand for copper resulting in price increases or falling supplies of 2 per cent ores, might, given the available technology, transform the 0.4 per cent grade copper from its status as just part of the world stock of copper to that of a reserve. Even then the state of demand for the lower grade disseminated copper ores might be limited to the extent that only the sources of supply most accessible to the major markets would either come into production, or be considered for it, and thus be included as reserves. Later, in circumstances of further copper price rises, the more remote disseminated ores could be worked since this might now occur at reasonable cost. However, any sudden discovery of new major sources of copper at no higher grade but closer to major markets and sufficient to meet short- and medium-term demands could then take out these remote production points again, thus allowing them to revert to the status of a resource.

## 2.2 Mineral resources and reserves – some theoretical considerations

The complexities of the relationship between attempts to define the known resource base of a given mineral and the dynamic relationship between this and reserves of a mineral, which by definition presuppose that they are exploitable both in terms of known technology and at a price which the market is prepared to pay, can be demonstrated diagrammatically (Zwartendyk 1972). In Fig. 2.1 the 'black box' (marked 1) represents the reserves of any given mineral. The size of these reserves may be increased by the discovery of hitherto unknown deposits of the mineral which are immediately exploitable given extant technology and prices (shown on the right), or added to or subtracted from by changes in the market price of the mineral or extraction costs. For example, a rise in the price of a mineral may mean that a deposit which was considered sub-economic and part of the resource base, can now be worked at a profit and thus considered as a reserve. A new less-costly mining technique or ore recovery process could have a similar effect

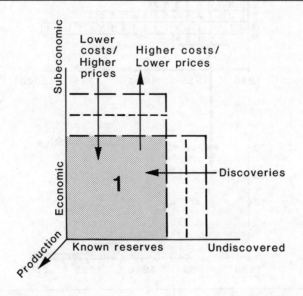

*Fig. 2.1*  The 'reserve' dynamic. *Source*: Adapted from Shank (1980)

transferring part of the resource base to the reserves. Conversely what was, at the then prevailing prices, part of the known reserves could move back to the status of being part of the resource base if prices fell as a result of a drop in demand, or if, in a given country, new higher costs were imposed on mining by the imposition of taxes, higher charges for energy used in processing, or the more rigorous adherence to higher standards of environmental control during the extraction of the mineral, its processing and/or during the subsequent period of site rehabilitation. This situation is represented by the arrows in and out of the 'black box' at the top of the diagram. Of course, demands are constantly being made on the reserves 'black box' by the consumers of minerals. Ultimately, if demands on the reserves of the mineral cannot be met by new discoveries which can be exploited at prevailing prices, the 'black box' will be augmented by drawing down sub-economic resources, unless, of course, substitute materials are found, or more of the minerals which ultimately find their way into finished products are recycled.

One way in which a mining nation or those concerned with the availability of world-wide supplies may monitor the continuing availability of its reserves of a given mineral is demonstrated in Fig. 2.2(a). The number of mines working a given mineral may be recorded graphically on the vertical axis

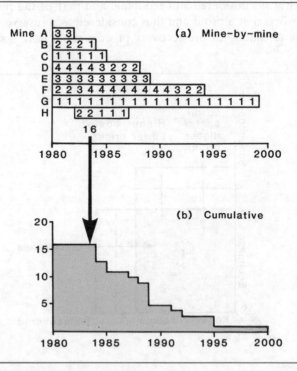

*Fig. 2.2* (a) (b) Monitoring reserves. Mine numbers are in tens of thousands of tonnes. *Source*: Adapted from Shank (1980)

whilst the life of each of these can be shown along the horizontal axis. Within these 'life' bars the expected output per annum can be recorded; the tonnages shown in Fig. 2.2(a) might represent in units of 10,000 tonnes the amount of metal concentrates recoverable from the ore body. Thus mine D during 1980 produced 40,000 tonnes, an output which would fall to 20,000 tonnes in 1986 and remain at this level until closure in 1988 when current reserves are expected to be exhausted.

A monitoring curve Fig. 2.2(b) shows each year's production from all the mines, thus representing a projection of total mine output year by year in terms of metal concentrates that may be obtained from currently known reserves. For example, for the year 1983 the total production of all mines is 160,000 tonnes. The curve thus shown demonstrates a pre-eminence of sustained levels of output in the medium term, but a gradual decline in the long term with only a few mines left with reserves of any size. Thus if the original 'black box' (Fig. 2.1) were a country's available reserves of a given metal, the curve would represent the rate of depletion if no new reserves were added to it from the resource base by changing mining cost factors or by the discovery of fresh resources which could be translated into reserves. The expected rate of mineral production from current reserves may be compared with expected levels of demand by plotting a demand curve on the same graph as the monitoring curve (Fig. 2.3). The widening gap between the two

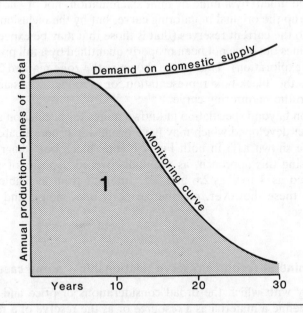

*Fig. 2.3* Expected rate of mineral production from reserves in relation to demand. *Source*: Adapted from Shank (1980)

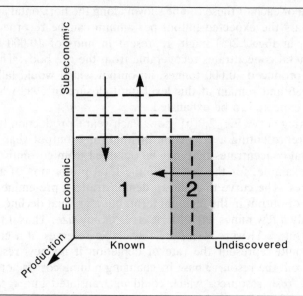

*Fig. 2.4* Inferred extensions to reserves (2) at developed properties. *Source*: Adapted from Shank (1980)

curves can be 'filled' by a mine by mine examination, not of *known* reserves which make up the original monitoring curve, but by the inclusion of inferred extensions to the current reserves (that is those that may be expected to exist at known mines but have not been properly quantified by a full programme of drilling and exploration). This can be shown as the zone marked 2 in Fig. 2.4 which shows the 'black box' representation, or as in Fig. 2.5 as an additional increment to the monitoring curve.

Moving on beyond speculation at known mines to speculation about properties not yet developed which may have promising if unevaluated deposits, these can be shown as 3 in both Fig. 2.6 (the 'black box' diagram) and in Fig. 2.7. Using this approach, as yet undiscovered deposits of the mineral can be added as 4 to Figs 2.6 and 2.7. Inferred from available geological information these, however, may be currently uneconomic and part of the resource base.

## 2.3 The mineral resource/reserve relationship – some case studies

The subtlety with which the broad considerations of price and technology interact to define a material as a resource or as the reserve of a resource can best be illustrated by some case studies. Close by the china clay workings of Lee Moor which formed the basis of the introduction in Chapter 1, is the

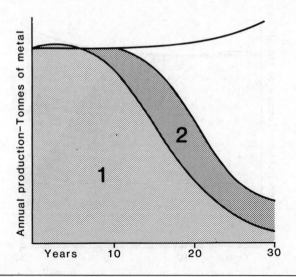

*Fig. 2.5* Inferred extensions to reserves (2) at developed properties. *Source*: Adapted from Shank (1980)

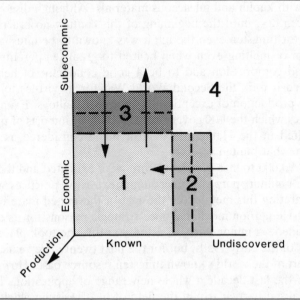

*Fig. 2.6* Possible extensions to reserves at unevaluated properties (3) or undiscovered properties (4). *Source*: Adapted from Shank (1980)

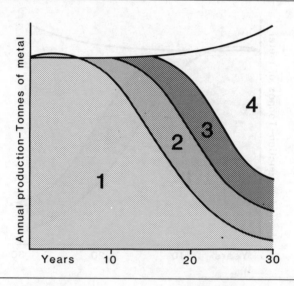

*Fig. 2.7* Possible extensions to reserves at unevaluated properties (3) or undiscovered properties (4). *Source*: Adapted from Shank (1980)

hamlet of Hemerdon. A disseminated ore deposit has been found there which mainly consists of tungsten (wolframite) but with some tin (casserite) intermixed with kaolin and micaceous materials. Although discovered in the eighteenth century, until the beginning of this century no practical use had been found for tungsten even though it was known to be unusually hard, to show no sign of melting even when heated to great temperatures, to resist corrosion and compression and to be a good conductor of heat and electricity. Indeed it took the Second World War for its virtues to be fully realised in the production of exceptionally tough steel alloys. Even then it was the difficulties which the UK government faced over imports of tungsten that at last resulted in the Hemerdon deposit being considered as part of the nation's somewhat limited reserves.

Although 90,000 tonnes of valuable ore were extracted and the metal produced used for military purposes, uneconomic extractive practices and the problems of separating this metal from the rest of the mined material made the shutdown of Hemerdon inevitable once strategic reasons for its existence no longer prevailed. A minor flurry of activity at this site took place during the Korean War of the early 1950s, but for the next twenty-five years this deposit remained part of the world's known tungsten resource base. However, by the beginning of the last decade a whole new range of applications for tungsten had been realised particularly in the field of heat resistant shields for space vehicles, for the manufacture of leading wing edges on supersonic aircraft, as radiation shielding for X ray and cathode ray tubes and in the manufacture of semi-conductor circuits. All these and other applications greatly increased world demand for the metal. Moreover, new lower cost open-pit techniques had become an established part of mining practice and ways had been found of cheaply and efficiently removing a large percentage of the tungsten ore from the casserite/kaolin/mica matrix of this deposit. Thus by the end of the 1970s, Hemerdon tungsten was again part of the total workable reserves of tungsten ore, over half of which is located in the USSR, China and North Korea. With a potential mine production evaluated at over $7.87 million (1980 prices) from a likely output of 54.5 million kg of ore concentrates over twenty years, it has become the largest reserve of tungsten in Western Europe and promises to be the fifth most productive tungsten working in the world by the time mining commences in the mid-1980s.

One aspect of the account of tungsten at Hemerdon underlines the impact of the intervention of an artificial constraint to turn what was a part of the total resource stock into part of the reserves of that mineral – that of war and a government's response to it. But governments and other agencies can intervene in other ways to promote a mineral to the status of a reserve or vice versa. Zimmermann (1951), the pioneer resource geographer, cited the case of the iron ore of Minas Gerais in Brazil, known for many years to be a major deposit of this mineral resource. The ore was, however, not substantially worked because of its inaccessible location far inland and a lack of suitable nearby coking coal to smelt it.

After the Second World War, the United States 'Good Neighbor Policy' brought with it, in the case of Brazil, both loans and expert advice on the construction of a modern steel mill. As a result a steelworks was built in a swampy, unhealthy river valley running inland from Rio de Janeiro, and the railway from it to the port of Victoria was refurbished. Thus, as Zimmermann (1951), stated, 'the stage was set for Brazil's iron deposits', for long merely part of the total iron resource base, 'to become part of Brazil's iron reserve'. As he pointed out, the conversion of these deposits to the status of reserves involved six major factors: (i) the co-operation of the United States government; (ii) capital in the form of credit, equipment and know-how; (iii) labour willing and able to work the iron ore; (iv) a domestic market for steel products, subsidised against competition from foreign goods; (v) a capacity to create an acceptable environment for those working in the steel plants; and (vi) the introduction of modern technology of steel production.

The influence of government in the Brazilian example was clearly crucial. Since then this phenomenon has become increasingly apparent, not just in centrally planned economies such as those of the USSR and its satellite states, but also in mixed economies such as that of the UK. In the case of the latter, the balance between reserve and resource status can be substantially influenced by environmental and taxation considerations, a matter which will examined in greater detail in Chapter 4.

The question of clearly distinguishing between resources and reserves is far from being a purely academic exercise. Without such a distinction, the crucial question regarding the adequacy of supplies of certain minerals, whether they concern a single nation, a wider economic community such as the EEC, or the world itself cannot be addressed.

As Van Rensburg (1975) pointed out with reference to South Africa, a number of estimates of the availability of indigenous supplies of coal (Wybergh, 1928; Venter 1952; De Villiers 1959) were based entirely on estimates of coal *in situ*. More recent studies deploying the same methodology including one as late as 1967, according to Van Rensburg had put the amount of available coal at $80 \times 10^9$ tonnes. These figures when married to annual demand considerations ($40 \times 10^6$ tonnes) suggested coal availability would be adequate for about 2,000 years. But such simplistic thinking has been subsequently considered by the Coal Advisory Board of South Africa as a quite inadequate basis on which to undertake forward planning in terms of either a development or a conservation policy for this mineral. The Board therefore recalculated the likely availability of coal, taking account of crucial concepts which distinguished between: (a) reserves (including potential reserves) on the one hand and resources on the other; (b) resources in commerical collieries in fully prospected areas and those in incompletely prospected areas; (c) *in situ*, extractable and saleable reserves in contrast to the simpler notion of reserves and potential reserves; and (d) various types of coal (recognising the important of different markets). The Board also recognised that (e) extractable reserve totals could change with different mining methods; and (f) most importantly, reserves are also a function of markets and uses. The results of this new work indicated that previous reserves estimates were over-optimistic; that they did not recognise prevailing economic factors limited the amount of coal which can be recovered at any given time; that they did not appreciate the relationship between grades of coal and market needs; that they did not recognise the meaning of reserves in relation to present and anticipated demand patterns for coal by various markets; that they did not recognise economic and technological factors determined the amount of coal recovered *in situ* and hence the magnitude of saleable reserves; that they did not recognise the demand for coal was essentially dynamic and not static; and that they did not appreciate coal was only one element in the energy economy.

This approach used by the Coal Advisory Board in the identification of reserves can be applied, with modifications, to other minerals. The important point is that the study by the Board illustrates the extreme complexity of estimating reserves, the dynamic nature of reserves in relation to resources, and when taken with the other case studies, the dependence of that relationship on political and social factors which in turn relate to demand.

# CHAPTER 3

# Classifying mineral resources

## 3.1 Resource categorization

Mineral resources are, of course only one part of that broad sweep of resources that are usually referred to as natural resources. Of the numerous attempts geographers have made to classify these, the most commonly one used distinguishes three main categories (Fig. 3.1). First there are *flow resources*, including all those items that may be depleted, sustained or increased by the actions of man. An example would be forests. Although these occur naturally, they may, of course, be destroyed in part or entirely removed as in the case of areas of the equatorial rain forests of Brazil. On the other hand, they may be originated or sustained by man. The activities of the Forestry Commission in the UK through its management of woodland on behalf of the nation provide a good example. Second, there are *continuous resources*, consisting of a range of resources which are always available independent of man's actions as well as those that are always available yet are capable of modification by man. Of the former, resources of solar and tidal energy are good examples, though, of course, the *use* made of them by man is very much under his control. The latter are exemplified by such resources as amenity landscapes. Whilst these often appear to be natural, man frequently has played a major role in their evolution. A National Park such as Dartmoor would be a good example where the natural climax of vegetation, indeed the landforms themselves have been modified by centuries of agricultural practice and mining, and more recently afforestation, water conservation and the development of recreation and tourism (Blunden 1977).

The final category is *stock resources*. These include all the minerals valued and utilised by man including aggregate materials (materials used in roads and building construction), other non-metallic minerals (for example, china clay, gypsum, salt, fluorspar, barium), metalliferous minerals including both ferrous (iron ore) and non-ferrous metals (copper, lead, tin and so on) and

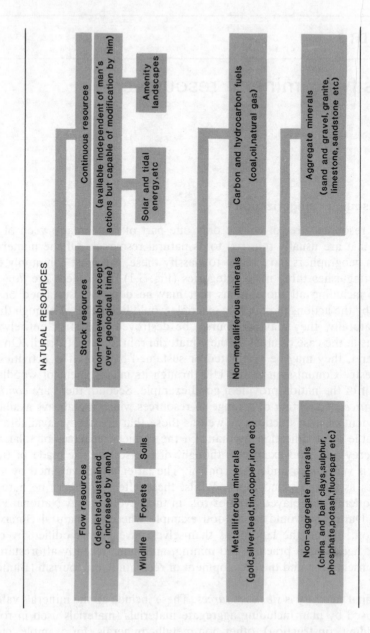

*Fig. 3.1* A resource categorisation. Source: Blunden (1977)

Table 3.1 Minerals: distribution, output, waste to ore ratios and environmental problems

| Class | Examples | Distribution and location | Ratio of ore/mineral to waste | World output measured in | Usual method of working | Possible environmental impacts* | Remarks |
|---|---|---|---|---|---|---|---|
| Common rocks | Limestone, chalk granite, sandstone, slate | Widespread and abundant | Almost all used (a) | Thousands of million tonne | Quarrying on surface of hillside | Scenic scars, loss of habitats but interesting when worked out | (a) Except for up to 95 per cent wastage in slate |
| Common rocks | Sand and gravel | do. | do. | do. | Wet or dry surface pits | Voids; flooding; lowered surface levels; drainage (f) | (f) Sometimes creating new water habitats and recreation areas |
| Earths and clays | Ball clay, stoneware clay, china clay, fullers' earth | Common | All used except for overburden (e) | Hundreds of million tonne | Surface working | Lowered surface levels; drainage problems; pollution; tips from china clay | (e) except for china clay where high percentage of waste material (1:10) |
| Common rock-forming minerals | Felspar, dolomite mica, quartz, fluorspar. | Common but (b) limited source | 1:1 down to 1:50 | Million tonne | Surface or hillside quarrying | Scenic scars | (b) need selective quarrying |
| Precious and semi-precious stones | Diamonds, rubies, sapphires, emeralds, opal, garnet, amethyst, jade | Rare | Stones are a minute percentage of waste rock | Thousands of grammes (kilogrammes) | Open pit (c) sands and gravels; underground mining | Voids and scenic scars | (c) or found in working other minerals |

| Class | Examples | Distribution and location | Ratio of ore/mineral to waste | World output measured in | Usual method of working | Possible environmental impacts* | Remarks |
|---|---|---|---|---|---|---|---|
| Common minerals | Asbestos vermiculite, pyrites, talc, soapstone, alum, barium, gypsum | Fairly abundant in limited locations | | Million tonne | Quarrying on surface or hillside | (as rocks): risks of pollution to water | |
| Less common minerals | Graphite, sillimanite wollastonite, Cryolite | Infrequently found | | Hundreds of thousand tonne | Adit mining into veins and dykes | (as rocks): risks of pollution to water | |
| Salts | Salt, rock salt, sodium salts, borax, potash, nitrate, phosphate | Common and fairly abundant in limited locations | | Million tonne | Deep mining, surface quarries; alluvial mining and solution mining | Waste heaps: subsidence; saline flashes | |
| Abrasives | Corundum, emery, pumice, commercial garnet | Rarely located but abundant when found | | Million tonne | Surface working of outcrop and adits into veins | (as rocks) | |
| Common metal ores (ferrous) | Magnetite, haematite, limonite | Abundant but in limited localities | | Hundreds of million tonne | Deep mining. Pillar and stall longwall; drift mines; opencast | Waste deposits; hill and vale restoration | |

| | Examples | Distribution | Ore content | Quantity / output | Mining method | Environmental impacts | Notes |
|---|---|---|---|---|---|---|---|
| Common metal ores (non-ferrous) | Bauxite, galena, nickel ores, tin ores, copper ores, zinc | Limited distribution | | Million tonne | Historically by deep mining, now mainly open pit: some alluvial (tin) | Voids; waste heaps; tailing dams and lakes; polluted run-off; toxic wastes | |
| Less-common metals | Manganese, antimony, cadmium, chromium, cobalt, mercury | (d) | | Hundreds of thousand tonne | Deep mining mainly by adits; some open pit | Toxic wastes | (d) most are only got as a by product |
| Rare metals | Indium germanium, lithium, caesium, selenium, tellurium, tungsten, thorium, titanium, uranium, vanadium, zirconium | Rare to very rare | 1:100 down to 1:5,000,000 | Ore output varies from under a tonne to thousand tonnes, according to scarceness and demand. Metals usually measured in 1000 lb. ounces (kilogrammes) | Various (d) | Toxic wastes, radiation and similar risks | (d) do. |
| 'Noble' metals | Gold, silver, platinum, palladium | Very rare | Ore contains about 0.1 per cent metal | Million troy ounces, (kilogrammes) | Deep mines with shafts and galleries or drift mines or alluvial mining (g) | Voids; waste heaps; scenic scars | (g) or as by products |

# Classifying mineral resources

| Class | Examples | Distribution and location | Ratio of ore/mineral to waste | World output measured in | Usual method of working | Possible environmental impacts* | Remarks |
|---|---|---|---|---|---|---|---|
| Fossil fuels | Coal | Fairly common | 2:1 (deep mining) 1:15 (opencast) | Hundreds of million tonne | Deep mines, with shafts and galleries or drift mines opencast, strip and auger mines | Subsidence; shale tips; scenic damage; air pollution from burning tips; liquid effluent; pollutants; temp. scenic damage by opencast | |
| | Oil | Abundant but rarely found | All used either crude, refined or in by products | Hundreds of million tonne | Land or sea walls | Oil spillage at sea or from pipelines; spoil heaps from oil shale working | |
| | Peat | Common | All used | Million tonne | Surface working | Lowered land levels; drainage problems; may destroy or preserve bog habitats | |

*Destruction of surface ecology is implicit in most classes

Source: Based on Lovejoy (1973).

Note: The simple classification used here follows that of D. Lovejoy and is not used in this text, although all the minerals of a major industrial significance listed are subsequently discussed. Those that are omitted from consideration in Chapters 5, 6 and 7 are given below along with their main uses with the exception of precious and semi-precious stones.

Fuller's earth: foundry moulds, fillers, cosmetics, lubricants, detergents, ceramics, drilling mud in oil well drilling operations, sugar and glucose refining, glyceride oil refining and soap manufacture.

Felspar: glass making, ceramics, cement, enamels.

Mica: insulation.

Quartz: glass making, refractories.

Vermiculite: insulation, light-weight plasters, lubrication.

Talc: lubrication, insulation, filler for paint and paper.

Cryolite: dyeing.

Graphite: electrodes, generator brushes, lubricants, pencils.

Sillimanite: refractory porcelain and bricks.

Wollastonite: ceramics, fillers, soil conditioners.

Nitrate: fertilisers.

Germanium: alloys.

Caesium: salts, rare alkalies.

Thorium: alloys.

Zirconium: refractories, pigments.

the carbon and hydro-carbon fuels (coal, oil and natural gas). Table 3.1 gives an overview of the full range of mineral resources involved, the detail of which is filled out in the later chapters. The table summarises some of the main points concerning minerals distribution, the quantities worked, amounts of waste involved in winning the valuable material, methods of working and environmental impact. Since all minerals in this stock category are only renewable over geological time (that is millions of years), they may be considered finite and theoretically can be used up. The chances of this occurring varies according to which of the broad categories of minerals is examined.

## 3.2 Stock resources and their availability

Metallic minerals exist at some percentage in all rock formations. One way to compare the total relative abundance of different metallic ores is to assume the earth to be a homogenous mass with all metallic ores distributed evenly through it. On this basis it has been calculated that a cubic kilometre of rock would contain 239 million tonnes of aluminium, 149 million tonnes of iron, 62 million tonnes of magnesium and 2,850,000 tonnes of manganese. These are relatively plentiful metals. At the other end of the scale a cubic kilometre of rock would yield about 239,000 tonnes of zinc, 155,000 tonnes of copper, 44,000 tonnes of lead and 14 tonnes of gold (Blunden 1977). But such figures relating to a theoretical average abundance throughout the earth are not related, of course, to the reality of extractive activity where minerals such as copper are likely to be perceived as reserves only at concentrations something like 1,000 times such average crustal figures (or in the case of lead, 3,000 times). In other words, the extraction of minerals often occurs where a chain of geological processes has concentrated minerals at percentages far higher than their theoretical crustal average, usually in the form of veins of mineral running through other rock formations, or where these veins have been exposed and eroded, as metallic ores disseminated through other sedimentary materials though at very much higher than average earth levels.

Working the scarcer metalliferous minerals at levels which approach their average abundance in the earth's crust, as a result of shortage, is a matter for the longer term and is discussed in greater detail in Chapter 11. More immediate concern regarding overall shortages of such minerals has been widely publicised and is based on the notion that the mining of deposits has in recent years occurred at a faster rate than the identification of new ones which have in any case been more costly to discover in relation of their likely output. In attempting to shed light on the substance of these contentions, Cranstone (1980), has examined 800 deposits covering a wide range of minerals discovered between 1946 and 1978. All of these deposits were deemed economically attractive enough to justify the expenditure needed to establish the grade of the mineral and likely tonnages available. Although

his work was confined to Canada, there is no reason to believe that the conclusions reached do not have a wider applicability. The investigations by
Cranstone (1980) have revealed that for more recent discoveries little information has been made publicly available which might assist in a more accurate measurement of the real state of exploration activity and also that
many of the sites evaluated, though not offering any immediate potential as
economically viable workings, will ultimately become so. Thus they will
move from resource to reserve status. In re-examining area-related work
previously carried out into the listing of new ore discoveries, on the basis
of more up-to-date evidence he noted that in his original study (Cranstone
and Martin 1973), 63 discoveries for the period 1966–73 should have been
recorded, rather than 37. In considering the period 1965–76, similar work
by Derry and Booth (1978) had listed only 52 when the figure should have
been 120.

As far as estimates of the value of the deposits is concerned (that is the
metals contained in the ore multiplied by metal prices), Cranstone has shown
that these have consistently erred on the conservative side, a situation only
partly explained by the propensity at the time of the discovery, or even after
the start of production, to underestimate the amount of ore likely to be recovered. At the same time the value of metals discovered per dollar spent

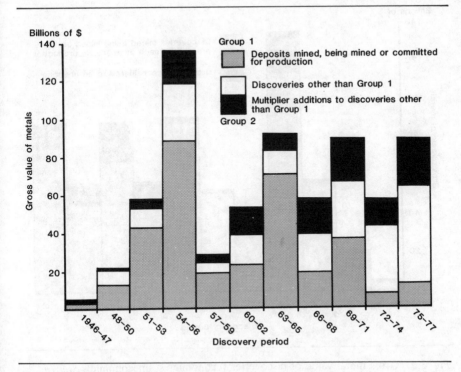

Fig. 3.2   Gross metal value of discoveries. *Source*: Cranstone (1980)

on exploration has not fallen in Canada, contrary to belief, remaining steady over the period 1946 to 1977 with the notable exception of the late 1940s and early 1950s when geophysical methods were introduced replacing conventional prospecting techniques (Fig. 3.2).

Finally, the contention that minerals are being used faster than new ones are being found must be rejected following a comparative evaluation of Figs 3.3 with 3.4 and 3.5. All are calculated for the same minerals, three-year periods and places, and are plotted on identical scales. This comparison shows that since the beginning of 1948 discoveries have substantially exceeded production for every three-year period except the two most recent. From 1972 on it is not lack of suitable discoveries but of hesitation by the industry to develop discoveries into mines that has brought this situation about. The factors which have caused this more recent pattern include unfavourable market conditions, taxation, more stringent environmental controls and so on. These are discussed in Chapter 4 in which factors operating on the short-term supply of and demand for minerals are considered.

As for the carbon and hydrocarbons (that is the stock energy resources) it has to be remembered that there is a relationship between levels of industrialisation and energy consumption. It is not surprising to find the USA

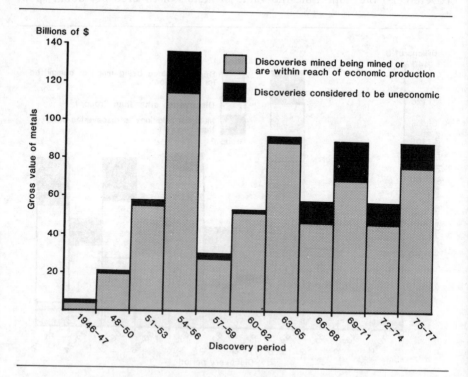

*Fig. 3.3* Gross metal value of discoveries (economic vs. uneconomic). *Source*: Cranstone (1980)

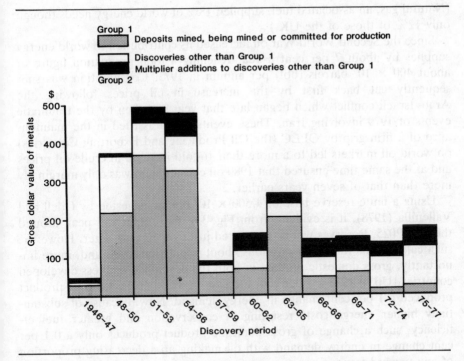

**Group 1**
Deposits mined, being mined or committed for production

Discoveries other than Group 1
Multiplier additions to discoveries other than Group 1
**Group 2**

*Fig. 3.4* Discovery value per exploration dollar. *Source*: Cranstone (1980)

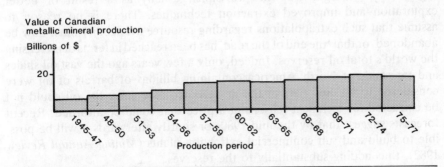

*Fig. 3.5* Value of Canadian metallic mineral production. *Source*: Cranstone (1980)

before the energy crises of the 1970s was consuming double per capita that of the UK whose consumption was five times that of most of the less developed countries. Oil is of greatest significance amongst the fossil fuels in terms of energy supply. In 1980 oil provided 43 per cent of world energy needs and in the UK, a typical industrially developed country, 48 per cent.

(Natural gas, an associated fuel, supplied 18% of world energy needs though only 12% of those of the UK.)

Since the Second World War oil increased its contribution to world energy supplies by about 7 per cent a year reaching a total production figure of about $400 \times 10^9$ barrels (bbl) per annum in 1973. Consumption was subsequently cut back first by the increase in oil prices following the Arab–Israeli conflict which began late that year and then by the traumatic events of 1979 involving Iran. These events which assisted in the maintenance of a firm grip by OPEC (the Oil Producing and Exporting Countries) on world oil markets led to a more than 16-fold increase in crude oil prices and at the same time ensured that 1980 oil consumption was only marginally more than that of seven years earlier.

Using a finite reserve base of $4,500 \times 10^9$ bbl as suggested by Odell and Vallenilla (1978), it is evident, from Fig. 3.6 that output will peak around the year 2025. Reserves will be exhausted just over 50 years later. However, such calculations reflect assumptions about population levels and, more importantly, gross domestic product in both the developed and less developed countries. Up to 1973 a 1 per cent increase in world gross domestic product produced a 1 per cent change in energy demand. Now, because of substantially higher energy costs resulting in conservation and greater fuel efficiency, such a change of gross domestic product produces only a 0.1 per cent change in energy demand with oil making up a decreasing proportion of any upward trend.

But there are other aspects of the question of future oil availability. As Fig. 3.7 indicates, since the 1940s, estimates of total resources have consistently undergone upward revisions, usually as a result of better exploration and improved extraction techniques. There is no reason to assume that such extrapolations regarding resource availability must now be abandoned, or that 'the end of the road' has been reached in terms of increasing the world's total oil reserves. Indeed, only a few years ago the vast oil shales and tar sands of North America containing billions of barrels of oil were considered to be only part of the oil resource base since the oil could not be extracted for technical reasons at a price that could be afforded. Recent forecasts suggest that by the mid-1980s or shortly afterwards it will be possible to build and run commercial extraction plants (*Mining Annual Review* 1982), thus adding substantially to the reserves.

Coal, the other mineral fuel presently utilised, is also a diminishing resource. This fossil fuel, on which the Industrial Revolution in Britain was founded, currently contributes 37 per cent to world energy needs or, in the case of the UK, 39 per cent. Until the oil crises in the 1970s, world production was rising at a steady 3.6 per cent per annum, though in the industrialised countries, especially in the USA, it was making a diminishing contribution to the energy mix compared with oil and natural gas. With global reserves thought to be around $7.6 \times 10^{12}$ tonnes and output for 1980 at $2.9 \times 10^9$ tonnes, forecasts are that 80 per cent will have been consumed be-

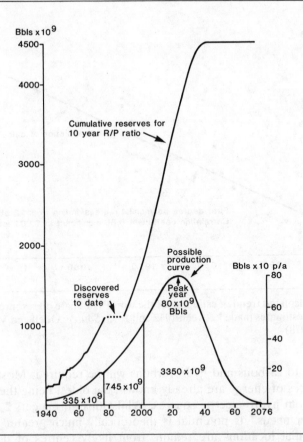

*Fig. 3.6* World oil reserves and production with 4500 × 10⁹ barrels oil resource
base and a rising demand curve varying over time from 4% to 3%.
335 × 10⁹ = cumulative production to end of 1978.
745 × 10⁹ = cumulative production to end of 2002.
*Source*: Odell and Vallenilla (1978)

tween the years 2040 and 2380 (*Mining Annual Review* 1982), but this calculation is as fraught with difficulties as that for oil, not least because there
are substantial deposits of coal that cannot be worked by known technology
and thus cannot be considered as reserves.

The upward trend in energy demand is inevitable although perhaps as
uncertain in quantitative terms as other forecasts. Whilst, as Fig. 3.8 shows,
the graph line indicating energy demand has levelled off since 1973 – with
the 1980 world total of 3,800 million tonnes of oil equivalent only marginally
higher than that of seven years earlier, and with current forecasts for the
year 2,000 between 5,000 and 5,800 million tonnes of oil equivalent, well
down on the 1977 prognostication by OECD of 9,000 millions – ultimately

41

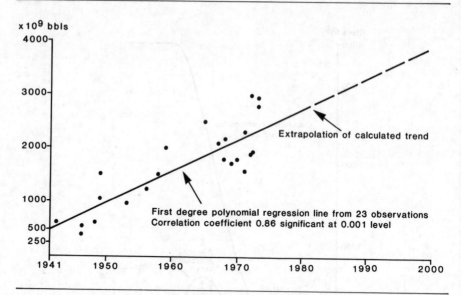

*Fig. 3.7* Calculated trend of estimates of the world's ultimate oil resources based on 23 estimates made between 1942 and 1975. *Source*: Odell and Vallenilla (1978)

alternatives to carbons and hydrocarbons will be required. Most of the potential sources of energy are already known, the oldest being the generation of power from hydro-electric sources. Whilst this only meets 2 per cent of world power needs, its potential is theoretically much greater except that many potential locations are remote from likely centres of demand. Although power can be carried long distances by cable, costs are high and losses of energy in transmission lines considerable.

Of the other possibilities solar energy appears attractive since there seems to be no effective limit to the amount of power that can be generated in this way. Energy can be trapped in solar cells, the output from which depends on the strength of the incident light and on the area exposed. Unfortunately, the amount of land in a country such as Britain that would have to be taken up with those cells in order to produce quite modest amounts of power make this mode of energy production unacceptable. A calculation which uses the average incidence of solar power at the earth's surface and assumes an energy conversion efficiency of 10 per cent indicates that to generate an output of 1,000 megawatts a site of 42 km$^2$ would be needed (Blunden 1977). Other more efficient means of trapping this energy source are being experimented with, including the use of orbiting solar panels, but nothing is in sight which could make feasible the generation of power using this means on an industrial scale.

Solar energy can also be transformed into other forms of energy in the

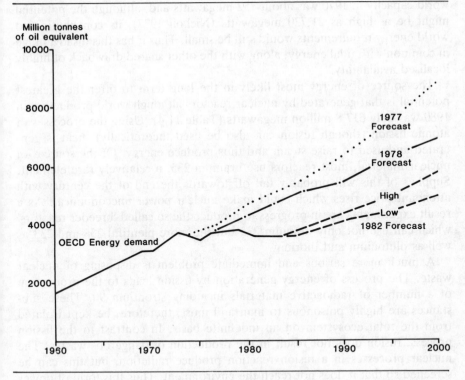

*Fig. 3.8* Energy demand forecasts. *Source*: International Energy Agency (1982)

atmosphere such as wind and, although only a very small amount is converted in this way, its theoretical potential is equal to 500 times the world's present energy consumption. Unfortunately the problems of harnessing such
transformed energy sources as a means of supplying our energy needs is not
far advanced except on an individual and localised scale. Indeed, it has been
calculated that using present technology 2,000 windmills, each the size of the
Scott monument, would be required to meet the city of Edinburgh's needs
(Nicholl 1977). On the other hand as Table 11.4 shows, the use of tidal
and geothermal sources of power at least offer some immediate prospect of an
input to energy supplies. A tidal scheme exists in France, others are planned
in Canada and Russia and, although relatively small, there is a theoretical
possibility of increasing world output to about 13,000 megawatts. Using heat
from the earth as a source of energy has long been practised in Iceland, Italy
and New Zealand and more recently in the USA, Japan and Mexico using
natural hot springs. It has also been experimented with in the UK where
heat-bearing rocks lie reasonably close to the surface of the earth. Here, as
at a limited range of other locations, such heat is capable of being exploited
by drilling wells and pumping down water as a heat exchange medium. Such
geothermal reservoirs can provide economic energy sources. However,

world capacity in 1970 was about 752 megawatts and, although the potential might be as high as 21,720 megawatts (Nicholl 1977), its contribution to world energy requirements would still be small. Thus it has this disadvantage in common with tidal energy, along with the other shared drawback of highly localised availability.

The source of energy most likely in the long term to offer the greatest potential is that generated by nuclear reactors although world production in 1980 was only 617.8 million megawatts (Table 11.4). Using the processes of atomic fission (though fusion can also be used theoretically), heat is generated and used to raise steam and thus produce energy. Of the sources of nuclear material, most reactors use uranium 235, a relatively rare element. Supplies of this will probably tail off towards the end of the century with attendant price rises which could make nuclear power uneconomical. As a result experiments are in progress to produce the so called 'breeder reactors' which can use not only uranium 235 but the more plentiful 238 and 233 as well as plutonium and thorium.

A much more serious and immediate problem is disposing of nuclear waste. The process of energy generation by fission leads to the production of a number of radioactive materials including strontium 90. These substances are highly poisonous to man and must, therefore, be kept isolated from the total ecosystem on an indefinite basis. In contrast to the fission process, fusion does not result in the production of dangerous wastes. The nuclear processes in a fusion reaction produce radiation, but this can be screened so that it does not reach the environment. Thus this form of power generation is far safer and, therefore, more acceptable in the longer run when located close to areas of high energy demand. For this reason efforts are being concentrated on the construction of fusion reactors of commercial viability. Forecasts are that these could become available before the end of the century. Although the potential for nuclear energy appears great, the rate of research and development of its productive capacity from the cheapest fuels and in the cleanest environmental form must be dependent not only on future demand but also on the availability and price of the major competing fuels to meet that demand.

Such energy considerations draw the discussion back towards the one other group of minerals that is yet to be addressed, the aggregate materials. The reason for this is that, while sand and gravel, limestone, sandstone and granite and so on exist in such vast quantities and are so widely distributed that the possibility of their total utilisation is remote, the working and processing of these materials could ultimately be affected by energy costs. For example, economic progress through industrialisation has been hampered already in some less developed countries by the fact that cement, the most important modern construction material which is manufactured mainly from limestone or chalk and involves the use of expensive hydrocarbons to fuel the kilns, can no longer be afforded. Indeed less developed countries without their own source of fuels, or inadequate foreign currency reserves with which to purchase them, are having to close down indigenous plant that was only constructed a few years ago.

## 3.3 Classifying stock resources

Although to undertake the previous analysis some rudimentary minerals classification was offered, the framework was too simplistic to provide any *insight* into the nature of what is a very broad range of mineral resources. If a classification does not help to provide insight, it misses much of its real value. The work of the geographers Jones and Darkenwald (1965) in classifying minerals on the basis of physical and chemical characteristics, as well as on the basis of end usage, has little utility to those concerned with spatial considerations. For this reason Blunden (1975) has devised a classification which introduces a vital distinction of a locational nature and draws upon the concept of place value. For this system five main categories are used: (i) ubiquitous non-metalliferous minerals: (ii) localised non-metalliferous minerals; (iii) ferrous metals; (iv) non-ferrous metals; (v) nuclear metals (vi) carbon and hydrocarbon fuels.

The broad distinction to be made between metals and non-metals derives from the differences in the physical characteristics of the two groups. Moreover the non-metallics are largely used in much the same form as extracted, require only a modest amount of processing, are mostly abundant and are generally of low cost compared with the metallics. Thus whilst aggregate materials may have cost as little as $2.5 per tonne (sand and gravel ex-works, 1980 USA prices), the equivalent per tonne prices for zinc, lead, copper and tin (London Metal Exchange) averaged respectively $760, $900, $2,180, $16,760. The prices of metal derive not only from their greater rarity of occurrence and the relative strength of demand for them, but also from the highly sophisticated processing needed to separate the metallic materials from the parent rock body. The amount of processing needed, whilst to some extent reflecting the chemical and physical properties of the ore body, is directly related to the concentration of the valuable mineral in the host material.

The ubiquitous non-metals groups are primarily utilised in the construction industry (for example, such aggregate materials as sand, gravel, limestone, sandstone, igneous and metamorphic rocks). Whilst many can substitute or interchange with one another, all have low unit values (undergoing relatively little processing) and high place values thus tending to be exploited close to the markets which they serve. Within the UK context, for example, prices for aggregates tend to double 50 km from their source when transport costs are included in the delivered price. The spatial distribution of such workings has therefore tended to reflect transport costs. Thus although quarry size has risen markedly in the last twenty years, with the costs of servicing wider market areas offset by economies of scale, improved transport and greater on-site processing so increasing the value of the product, such extractive operations still tend to be of medium size and show a dispersed pattern related to major centres of population.

Only at the top end of the ubiquitous non-metallics groups where brick clays have been transformed into bricks and limestone used for the production of cement, has the application of much more sophisticated processing led to the addition of substantial value to the basic raw material. This also has the effect of achieving a greater concentration of such processing enterprises serving larger market areas in spite of the wide availability of the basic raw material. But as Spooner (1981) has rightly pointed out in commenting on the classification devised by Blunden (1975) such enterprises 'contribute to the agglomeration of economic activity in a region rather than initiate new patterns of economic development'.

Compared with the ubiquitous non-metallics, the localised non-metallics have much higher unit values and include most of the raw materials for the chemicals and fertiliser industries. Examples are salt and potash, china clay and fluorspar. Although restricted in occurrence, the scale of production and the greater scarcity value of these minerals enable them to stand higher transport charges to more distant markets. Here the china clay deposits of the Hensbarrow district of Cornwall are a particularly good example of large scale highly rationalised mining development well placed adjacent to the coast to take advantage of low cost sea transport to overseas customers. Whilst no immediate infrastructural benefits have accrued to this part of Cornwall by way of local paper making and pottery manufacture in the way that the working of salt in Cheshire has led to regional industrial development based on that mineral, it would be misleading to assume that its value to the locality is negligible. Chapter 10 on mining and regional development offers evidence of the more than modest multiplier effect that those mineral workings which generate few 'down-stream' activities can actually have.

Apart from the distinction between localised and ubiquitous non-metallic minerals, another significant aspect of this classification is the separate identification of mineral fuels. These differ from all the other groupings primarily because they are not minerals which for the most part directly become part of an end product, but are consumed to provide motive power for industrial and domestic uses. Indeed fuel costs can have an important effect on the working of other minerals as will be seen in Chapters 5 and 6. The exception to what has been said is, of course, where carbon and hydrocarbons are used as stock feed for the petro-chemical industry. The influence of both fuel forms on regional development is also made abundantly clear in Chapter 10. Indeed the impact of North Sea oil on north-east Scotland and particularly the Shetland Islands is one of the major case studies.

Of the metallic minerals a distinction has been made in this classification between ferrous and non-ferrous metals. One reason for this is to draw attention to the fundamental role played by iron ore in the ultimate production of steel, a product which in its many forms lies at the heart of modern industrial economies. Perhaps of more significance for the split classification of metallic ores is the much greater ubiquity of iron ore compared with other metallics. A prime reason for this is that iron as a metal is derived from such a multiplicity of geological circumstances. Certainly ores have been extracted on a large scale from areas where the processes of geological concentration have occurred to form massive deposits of magnetite in pre-Cambrian rocks – those from Kiruna in Sweden are a good example. In contrast deposits laid down on the floor of shallow seas have given rise to the haematite workings of Krivoi Rog in the USSR, now the most important production area in the world. Other sources of iron particularly in Western Europe derive from the Jurassic ore fields where the principal component materials are iron carbonates and hydrated oxides of iron.

The wide geographical spread of iron ore production (as Fig. 6.1 emphasises) and its high metal content (the UK's average of 28% would be considered low by world standards), compare sharply with the far more scattered nature of the rest of the metallic ores, the distributive inequality of their deposits and their generally low grade (frequently below 1%). Nickel, for example, has its two most important sources located in a single country, Canada, where the deposits lie in the pre-Cambrian rocks of Ontario and Manitoba. The rest of current reserves (over 50%) are shared between Cuba and New Caledonia. The world's major molybdenum deposits exist on the eastern edge of the Rocky Mountains near Climax, Colorado, where the mines provide over half of the global total output. Most of the world's sources of manganese, a vital commodity in the production of steel, are in the USSR which produced only 20 per cent of world steel output in 1980. Tungsten and cobalt reserves are principally located in China and Zaire respectively, while a narrow belt running from southern China through Thailand, Malaysia and Indonesia has most of the world's tin. Bauxite is found primarily in the Caribbean while Mexico and Peru contain most of the known supplies of silver, and South Africa most of the world's gold. Even in the case of copper, Chile, Peru, Zambia and Zaire dominate in terms of known workable deposits.

Such inequalities in the distribution of the metallic ores have given rise to political difficulties. For example, the USSR as an act of deliberate policy withdrew its supplies of manganese to the USA in the late 1940s. As for oil, the impact of the changes in the supply and price of oil in the 1970s, was substantial on the industrialised countries of the world. The industrialised countries were unable to control either in the face of the producing countries acting together within OPEC as a corporate body.

But perhaps of greater long-term significance are the inequalities in patterns of consumption. The major industrialised nations currently consume the greatest quantities of minerals. As already noted the mineral ore, iron, lies at the centre of an industrial economy because it goes into so many products. If the way this is consumed around the world in the form of steel is

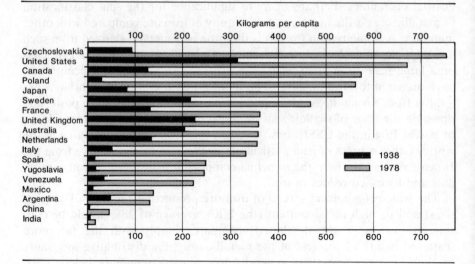

*Fig. 3.9* Steel consumption per head (kgs) 1938 and 1978. *Source: Metal Bulletin*

examined, the differences between the industrialised and non-industrialised nations immediately become apparent (Fig. 3.9).

Over the period 1938 to 1978 world steel consumption per head (excluding the USSR) rose by over 430 per cent. Within this increase, the highest rates were achieved by developing countries such as China where consumption increased by a factor of 15 times and in the Latin American countries of Columbia and Venezuela where the increases were of a factor of around nine and six respectively. Of the developed countries in Europe, those bordering the Mediterranean achieved the greatest increases. In Spain and Yugoslavia consumption per head increased 14-fold; in Turkey 8-fold; and in Greece, Italy and Portugal, some 6-fold. Whilst consumption per head more than doubled in the other older industrialised countries, only Japan achieved a spectacular increase with levels in 1978 six times greater than in 1938. In absolute terms per capita consumption in the USA has remained far and away the greatest over the period in question. In 1938 it was 314 kgs with its nearest rival the UK at 227 kgs. But by 1978 at 670 kgs it had just been overhauled by Czechoslovakia with Canada, Japan, the German Democratic Republic and the German Federal Republic close on its heels.

Thus, the world may be divided into four main groups of steel consumers. The first group consists of about 775 million people living in 22 industrialised countries who consumed steel in 1978 at rates varying between around 300–700 kgs per head (that is nearly 400 million tonnes all told in that year). The second group consists of around 345 million people living in 12 countries. These include some of the developed countries peripheral to the Western European industrial heartland. It also includes the larger developing

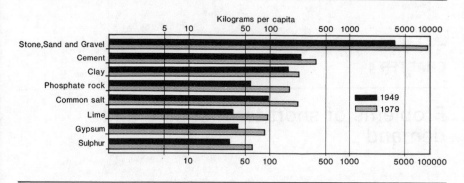

*Fig. 3.10* Building materials consumption per head (kgs) 1949 and 1979 in the USA.
*Source: Minerals Year Book*

nations of South America. In 1978 the countries in this group had a consumption per capita of 100–300 kgs (that is a total of 55 million tonnes of steel for all the countries in question). The third group is constituted by four developing nations with a per capita consumption of about 50–100 kgs and a total population of 1,064 millions of which China made up over 90 per cent. Total steel consumption in 1978 amounted to nearly 48 million tonnes. Of the rest of the world's steel users, the 684 million inhabitants of India consuming 16 kgs of steel per head in 1978 stand out from the remaining 1,400 million people not accounted for above and whose position in terms of steel consumption is even less favourable.

Although the problems of meeting such levels of consumption would not necessarily be so problematic for many of the non-metallic minerals, especially the commoner building materials, attempts to approach the per capita levels of consumption existing in the USA in 1979 (Fig. 3.10) by the less developed world would have a massive impact on the environment and would again have serious consequences with regard to levels of energy availability. The whole question of the future supply of and demand for minerals is dealt with at greater length in Chapter 11. As to current and short term questions, these are the subject of the next chapter.

# CHAPTER 4

# Problems of short-term supply and demand

## 4.1 Historical patterns of supply and demand

As the later chapters on individual minerals make clear, the current world distribution of these presents a substantially different pattern from that displayed by their use. As major producers of manufactured goods most Western European countries and Japan have long been the chief importers even though the former based its original industrialisation on indigenous supplies of iron ore and coal. In Great Britain, the beginnings of its Industrial Revolution saw it as the world's largest producer of metals so that there was a period in its history when the major producer and consumer was one and the same. This mantle was later taken over by the USA although it is now increasingly becoming an importer. Only the COMECON countries remain largely self-sufficient as an economic trading bloc, mainly as a result of what Adams (1980) has described as their 'policy predisposition autarky'. Even these nations, together with China, have begun to make small quantities of metalliferous minerals available to international markets, albeit on an irregular basis. This leaves the less developed countries in the role of prime suppliers to the advanced industrial nations. This situation applies across a wide range of metalliferous as well as other industrial minerals and, of course, to oil. Only the construction materials can be realistically excluded from this relationship.

In terms of world trade the relationship between suppliers and consumers has been a relatively dynamic one particularly over the last decade. Prior to that and certainly throughout the 1960s most metals were available to consuming nations in circumstances largely determined only by changes in production costs. In other words, the buyer could be relatively sure that given prices would remain in force over significant periods of time and that the market would maintain itself in an orderly and predictable fashion. The only exceptions to this situation could be observed with copper and lead which tended to be subject to short-term price fluctuations. For metallics such as aluminium and nickel, trading was arranged through specific consumer/supplier contracts. Although zinc was traded by the major free world

50

market organisation, the London Metal Exchange, a producer cartel covering a wide range of countries led to price stability. The same applied to nickel because it was mainly controlled by one nation, Canada. As for precious metals, the price of gold was fixed at $35 per oz. by the US government which also effectively controlled the price of silver. And, in the late 1960s when the price of the former was allowed for the first time to respond to market conditions, the premium over the official quotation exceeded it by as little as 20 per cent at its highest point.

## 4.2 Instability in the minerals market

The 1970s however, presented a different picture. The markets for both aluminium and nickel moved into a situation where consumer demand played a much more significant role with both metals handled on the London Metal Exchange. Certainly, what the free market was prepared to pay deviated markedly from what producers wished to sell at. The market for copper showed even greater instability. After an initial 400 per cent price rise followed by a decline to 33 per cent of the earlier levels, the trend of the last two years of the decade was generally downward but within a somewhat cyclical pattern. Zinc equally had a period of market price increase rising over 400 per cent up to 1973–74 before plunging to very low levels in 1976–77. The strain on producer-agreed prices was such that much less of this metal is now traded via such arrangements. Where tin is concerned, its trading had long taken place inside a price range controlled by a producer-dominated organisation, the International Tin Council. As through the 1970s the market price fell well outside the limitations agreed by this organisation, its value to its producer members became very limited and one may now consider it defunct.

The problems of pricing and supply which hit some of the more widely traded metals also had their impact on many of the important ferro-alloys described in Chapter 6. Chrome supplies were much affected by the political situation in Zimbabwe, whilst the price of cobalt rose substantially as a result of the war in Zaire in the late 1970s. Although the molybdenum supply has not been affected by political conditions in the less developed countries since the USA remains the major supplier, an adequate output of the mineral has not been steadily maintained in response to demand. Consequently prices have not remained predictable even in the short term.

The reasons for the instability in the minerals market have already been hinted at, but they deserve further general analysis to isolate political, financial and economic factors for specific consideration.

## 4.3 Political nationalism and mining investment

In political terms, a key issue of the 1970s was the gradual dissociation of decisions regarding mineral production and consumption. Prior to this period any general comment regarding the structure of the minerals industry could not have failed to underline the significance of the control of production by large multi-national companies many of whom exercised further market control by their ownership not only of minerals processing but also of primary fabrication. Moreover their command of output frequently exercised over mines in more than one country, allied with their market intelligence, ensured their hegemony in the market place. However, the rapid extension of territorial independence to former colonial countries, with its concomitant reduction in the foreign control of mining enterprises, sometimes through sudden expropriation by the newly independent government, meant much less co-ordination of supply and demand decisions. This situation was evident particularly in Latin America and Africa during the late 1960s. In the case of the central African countries alone, 1969 saw the Zambian government take a 50 per cent controlling interest in the two giant copper mining groups, Anglo-American Corporation and Roan Selection Trust, later rising to a 60 per cent plus holding of equity. In Zaire the government took over the interests of Union Miniere. In Botswana the state acquired an interest in De Beers Consolidated Mines' diamond venture and in the copper-nickel mines run by a consortium of foreign companies led by AMAX Inc.

The problems for the market were further underlined by the attitudes of both the multinationals, partly or wholly dispossessed of their mineral holdings, about future investment and the new nations themselves particularly concerning their own view of how they should exploit their natural resources. On the latter point, their own approach in no way resembled that of profit maximisation which characterised the multinationals. The need to pursue a policy of relatively full employment and to ensure the acquisition of usually much needed foreign exchange frequently led such nations to run production at a high level even though the world market did not always justify such an approach. For example, in the latter part of the last decade the down-turn in copper demand ideally should have been matched by a reduction in supply if prices were to be maintained at the levels of the immediately preceding years. However, the great dependence of less developed countries in Africa and South America on the mining of copper as their only source of revenue led them to continue their already high levels of production. These objectives then tended to exacerbate and prolong the down-turn in the price cycle.

Politicisation of the production side of the supply and demand equation may be exemplified in other ways since less developed countries have in some instances taken a less shortsighted view of their mineral resources as part of their national heritage which can be used to meet social, economic and political objectives. In the case of oil, the formation of OPEC and the attitudes of its member nations (certainly in its earlier years) demonstrate just this attitude as does that of the formation of the International Bauxite

Association, though in this case, its impact on consumer nations seems unlikely to remotely resemble that of OPEC with its achievements 'born of a unique geographical, political, institutional and technical parentage' (Manners 1978).

## 4.4 Environmental costs and mining development

Another major factor of a political nature which acted to destabilise the minerals market in the 1970s was that of environmental control. The impact of legislation to control dust, noise, effluent discharges to the natural drainage system, disposal of solid wastes as well as to ensure landscape rehabilitation in the case of mining enterprise, is discussed in Chapter 8. Here one needs only to note the particular importance of such legislation on mines in the developed countries and the significance of such measures of control as additional cost factors; what were externalities have become part of the internal cost structure. These added charges can be critical in the case of workings operating at the economic margins, whether they be metalliferous mines, aggregate undertakings or those otherwise connected with the construction industry.

For example, Papanicolaou and Mackenzie (1981) evaluated the effects of water pollution control on the economics of 131 base metal deposits discovered in Canada during the period 1951–74 as well as all other potential mining developments in that country. The deposits were mainly copper, zinc, lead and molybdenum. They found that total capital and operating costs (at 1979 prices) of the currently necessary pollution control measures ran out at $1,200,000,000 (Canadian). That figure is between 3 and 10 per cent of total mine investment and 2 and 5 per cent of operating costs, rising to $2,100,000,000 if it was assumed that the best possible water treatment schemes (given known technology) were employed. In the latter case water pollution controls would be 13 per cent of total investment and 8 per cent of operating costs.

As the report points out, while the impact of these controls varies according to mine size, deposit type and region, the incidence of expenditure to combat pollution is greatest for smaller sulphide deposits because of the need for more sophisticated means of treatment the costs of which cannot be spread across a large output of ore. However, given the prevailing market conditions for copper in 1979 and the rate of taxation, five deposits would be rendered uneconomic by the levels of water pollution control then required. This would have meant forgoing a potential net value to society (defined as the increase in society's real wealth by investing in minerals as opposed to some other activity) of around $700,000,000.

Mackenzie (1980) had earlier used 57 of the same sample of base metal deposits to consider the impact of air pollution ($SO_2$) resulting from smelting activities at their locations. This study was less detailed in its approach compared with the water pollution study. For example, emission control data from only one smelting company was used. Nevertheless the results show that control costs only marginally diminished the number of economic deposits. This is perhaps because of the location of the sample entirely on the Canadian Shield where deposits have strong profitability characteristics. However, assumed $SO_2$ control costs (based on legal requirements at the time of the study) did have a significant effect on the potential value to society of the base metal mining sector. When all other costs except taxation were taken into account, the amount by which pollution controls of this kind diminished their potential value was around $271,000,000.

There are specific tax advantages to be gained, however, from the use of integrated smelters which are presumed to be a part of the total processing operation (as opposed to the use of a separate 'custom' smelter) especially in Quebec and Ontario. This arrangement further favours the larger, higher ore grade mines. Overall these controls can be regarded as a disincentive to the development of smaller lower grade deposits while bringing extant medium and large lower grade mines closer to the margins of profitability. Clearly, either through reduced mining profits or tax reliefs on integrated smelters, the end result of air pollution controls for governments is a substantial reduction in their revenues from the minerals industry. But Mackenzie has calculated that if the levels of incentive for investment were to be the same as those which existed prior to the imposition of the $SO_2$ controls, a 13 per cent reduction in taxation levied by the national and provincial governments would be required.

If such environmental legislation affects the levels of supply, other recent legal requirements in developed countries can impact on minerals demand. The motor vehicle industry offers a particularly good example since alterations in the permissible levels of tetra-ethyl lead in petroleum in the US, Canada and more recently in EEC countries has caused considerable instability in this market. At the same time, a sudden demand for the platinum in catalytic convertors in exhaust systems designed to reduce air pollution had a similar impact. Moreover the construction of cars made from lighter materials such as plastics, as well as the lighter metals and metal alloys to meet higher levels of fuel efficiency required as a result of the great increase in petrol costs in the latter half of the last decade, has created further structural imbalances in the supply and demand of all such metals.

In short, the whole canon of environmental and other legislation has made it increasingly difficult for the supply system to react to changes in the market. Certainly such legislation seems to have been passed with little or no thought for its impact on the metals market. This view is clearly not without its support in some government circles since Ontario's Ministry of Natural Resources (1977) agreed in a report that changes in the standards of environmental controls had a distinct impact on mining investments in Ontario which is Canada's leading minerals province. This same report also argued cogently that a similar disincentive impact existed where tax changes were concerned and when combined with additional environmental burdens led to 'investment substantially lower than it would have been had the pre-1972 tax and environmental regimes prevailed'. Indeed in a subsequent study using a computerised simulation model, Anders *et al.* (1978), were able to show that 'the combined effect of changes (in taxation and environmental regulations) was a reduction in mineral investment of at least 20 per cent' over that which would have occurred in the province had these changes not been introduced.

## 4.5 Taxation and mining development

Turning to Canada as a whole, a detailed analysis of the effects of taxation has been undertaken by Mackenzie in collaboration with Bilodeau (1979). They identified 124 deposits in the base metal sector all discovered in the period 1950–74 but with widely varying characteristics and locations. Taking account of metal prices and the meeting of all capital costs (including exploration) considered in constant 1974 Canadian dollars but assuming the imposition of no taxes, 86 of the deposits proved to be economic offering a net potential value to society of $2,757,000,000. Using the 1969 Canadian mineral taxation structure as perhaps the most favourable to be yet applied so far as mining interests were concerned, they found that the number of economic deposits fell to 82 with a value of $2,646,000,000 (with taxation levied at a rate of 36%). Finally, using late 1976 tax levels 77 deposits could be assumed as economic offering a value to society of $2,484,000,000 (with taxation applied this time at 61%).

Canadian mining taxation structures also contain a provincial variant. Thus if 1976 is taken as a base line, Mackenzie and Bilodeau found that when the taxation system in Quebec was applied to the 86 deposits viable in a 'no tax' situation, 79 could be operated. Under the Ontario tax regime, the number would be 75; in British Columbia 74; and in Manitoba, 70. The net value to society of these mines would decrease from $2,548,000,000 in Quebec to $2,260,000,000 in Manitoba, with the net share for governments to be subtracted from these sums ranging from 64 per cent in British Columbia, to 65 per cent in Quebec, 72 per cent in Ontario and 83 per cent in Manitoba.

De Young (1977) also noted the crucial impact of the much less favourable Canadian taxation regimes (as far as the mining companies were concerned) following changes enacted in 1972 at national and provincial levels. He found that investment in mining as a percentage of total business investment decreased from 8.3 per cent in 1971 to around 3.9 per cent in 1975, whilst expenditure on exploration in Canada decreased at an annual average rate of 19.8 per cent between the year in which the major tax changes occurred and 1977. Indeed by 1975 Canadian mining companies were spending around 60 per cent of their exploration effort abroad, mainly working in the USA, South Africa and other African countries, compared with 20 per cent in 1971. As De Young (1977) concluded 'the tax environment presented by

55

a political jurisdiction is of particular importance because companies exploring for minerals are not constrained by provincial, state or national boundaries'.

## 4.6 Economic cycles and mining development

However, the tax burdens referred to in these studies were not those applied to excess profits of which the mining industry in North America and the United Kingdom has complained and which relate particularly to oil. These taxes have been levied on *windfall profits* where unexpected rises in the world price of oil have brought about a rapid rise in the value of the mineral assets of the oil companies through no endeavours of their own. Such profits have in the past, it is argued by the industry, been used to pay for the periods of unexpected losses. Such circumstances provide the margin of revenue which enables mining to survive the inevitable years of economic recession.

Cyclical trends in the economy have been and will always be a significant factor affecting the supply of and the demand for minerals. Where aggregates and building materials are concerned, demand tends to fall rapidly in line with the decline in new construction which always accompanies such downturns. As for metalliferous and other industrial minerals, a fall in demand for manufactured goods has a profound impact on the need for such basic inputs to the productive process. It was, of course, a mineral itself that sparked off the recession of the latter part of the 1970s and early 1980s, when in 1974–75 the spectacular four-fold rise in oil prices triggered an economic down-turn whose impact has been world-wide. Because of its spatial ubiquity and relative simultaneity, further enhanced by the second round of large oil price increases of 1979, this cyclical depression has been bigger and more completely synchronous in all parts of the world than those of the past.

## 4.7 Mining investment – decisions in an environment of uncertainty

Of the particular economic factors that created instability in the minerals market during the 1970s, the increasing size and complexity of new investment decisions must be considered a significant matter. For example, an integrated copper mine, smelter and refinery can involve an outlay of around $1 billion. Put another way, a new major copper mine with its processing plant and attendant infrastructure would cost around $8,000 per tonne of output in 1983, a rise of 3 to 4 times above the figure prevailing at the beginning of the 1970s. Certainly few metal mining projects anywhere would cost less than $100,000,000. But whilst the tungsten mine referred to in Chapter 10 spent upwards of $12,000,000 in completing its preliminary programme of exploration and testwork designed only to prove the viability of the deposit, it is in the extractive complexities of winning and processing such low grade ores where high levels of investment may particularly lie. The proposed tungsten operation involves open-pit working and the separation of the tungsten from a deposit which also contains china clay and tin.

Apart from the mining technicalities themselves, the problems of putting together a financial package to work the mine in such a way that will show a return on investment in an uncertain political environment are becoming increasingly difficult. A good example described by Barber (1980) is the Cuajone copper mine of southern Peru. Taking seven years to construct, it was preceded by several years of negotiations with the Peruvian government, the object of which was to provide an equitable division of the benefits of the mine for the mining company, the investors and the host country. Their outcome, ultimately consolidated in a 'Bilateral Agreement' of December 1969, acknowledged that the project 'forms a part of the national plan for maximum development of the country's copper production' and that 'the special guarantees of Article 56 of the Mining code will facilitate its financing abroad'. The Agreement set out in detail the benefits and guarantees afforded to the company and the foreign exchange tax regime to which it would be subject; that is 47.5% tax on net earnings during the period of investment recovery and 54.5% for the period of the life of the mine thereafter.

It took a further six years, well into the development period of the mine, to finalise the complex financing of the operation which included project loans from eleven separate suppliers of equipment in the UK, the USA and Brazil; Japanese and European copper buyers; the US Export–Import Bank, the World Bank, the International Finance Corporation, and a consortium of twenty-nine commercial banks with headquarters in eight different countries. Some of the supplier and copper buyer loans were supported by agencies of the lenders' government, including the Export Credit Guarantee Department in the UK. All these agencies contributed some 60 per cent of what was to be a joint project of the Southern Peru Copper Corporation (a consortium of four US mining companies) and Billiton N.V. and involving a total outlay of $726,000,000.

If all this were not enough complexity, other unforeseen factors subsequently entered into the production cost function of the mine to further complicate its relationship with the market. In 1971 a scheme was introduced to transfer the eventual ownership of Peruvian assets and earnings to the workers of mining companies through mining communities. This resulted in a 10 per cent charge against company earnings after tax had been paid. In the same year the government nationalised the export of mineral products from Peru and imposed a marketing fee equal to 2 per cent of the overseas sales value of the products to finance the new agency administering the scheme. This fee was in addition to the existing 2 per cent export tax applied to the mine's output, a figure raised by a further 1 per cent in 1979.

The result of these additional burdens can be seen either as factors eroding the competitive edge of Cuajone copper or as damaging the investor's expected rate of return on their investment. In the latter case, it would appear that the political requirements of the moment may have overridden longer term interests in that those conditions likely to attract future mining investment have been passed up. Thus, apart from the complexities of the relationship between a mining venture and its host government and the inevitably much longer lead time required between a decision to set up a mine and bring it into production (all adding to the complications of the supply/demand relationship) the Cuajone experience also shows that hard-pressed governments in less developed countries may well finance their social and other government programmes by extracting the revenue they need from mining companies which cannot resist and have no means of redress. It was this latter message that was certainly taken to heart during the late 1960s and 1970s by the multi-national mining corporations causing, by the beginning of the present decade, a dearth of minerals investment in the less developed countries. As the World Bank (1977) has shown, in the early 1970s only 15 per cent of exploration expenditure outside the centrally directed economies was incurred in the less developed countries while 80 per cent of all such expenditure was concentrated in just four countries of the developed world (Australia, Canada, South Africa and the USA). This situation was further underlined in 1978 at the 11th Commonwealth Mining and Metallurgical Congress when it was disclosed that exploration expenditure in the less developed countries by members of the Group of European Mining companies had fallen from about 34 per cent of their total expenditure in 1961–63 to 13.5 per cent in 1973–75.

## 4.8 Imbalance in the spatial patterns of mining investment

Unfortunately, the longer term repercussions of such a policy of investment could be catastrophic, not only in terms of a failure of the developed countries to generate economic growth in the less developed countries, but because it could ultimately lead to a severe shortage in supplies of those minerals on which the industrialised world depends. This can be demonstrated by a consideration of EEC countries alone in which the dependence on raw material imports is running at between 95 and 100 per cent for nickel, manganese, antimony, chromium, molybdenum, platinum, selenium, tantalum, titanium, tungsten, vanadium, zirconium, phosphate and asbestos. For aluminium, copper, cobalt, lead, tin, zinc, iron ore and uranium, the level of import dependence is between 50 and 80 per cent. Although sources of some of these materials are spread widely across the world, and supplies of many of these are forthcoming as a result of increased investment in mining ventures in Australia, Canada, South Africa and the USA, the significant point is that the less developed countries account for about 55 per cent of the total reserves. This highlights the ultimate inter-dependence between the present developed and less developed countries, a consideration which is evident again and again, particularly in Chapters 5 and 6, dealing with individual minerals, and underlines the need to stimulate mining investment in less developed countries if future demand levels are to be met.

The United Nations has shown its concern in this respect in a report (Gluschke, Shaw and Varon 1979) dealing with the investment requirements for the world copper industry between 1980 and 1990, and has come up with some instructive figures. Projected world consumption of refined copper is expected to grow from 9,936,500 tonnes in 1980 to around 16,460,000 tonnes in 1990, broken down as follows: developed countries from 6,553,000 to 9,885,000 tonnes; less developed countries from 748,000 to 2,124,000 tonnes; and centrally planned economies from 2,631,500 to 4,500,900 tonnnes. These projected levels of consumption assume rising production levels from new mines and recycled material, which in turn depends on the availability of investment to make that production possible. On the basis of its refined copper consumption and required productive capacity projections, the UN study suggests the investment requirements of the world copper industry (excluding the centrally planned economies) for mine, smelter and refinery capacity in the period 1977–90 at about $43,000,000,000 or $3000,000,000 a year. These are not only substantial sums of money which potential capital exporting countries must find in a period of economic recession, but sums which must inevitably in large part go to the less developed countries with potentially mineable copper ores and in spite of an investment reluctance stemming from those reasons already specified.

## 4.9 International aid and mining development in the less developed countries

One solution to the longer term problem of creating investment in the less developed countries has not been pursued through the normal mining finance channels but through the application of international aid. However the record here, whether with reference to governments or to international agencies, has so far not been very impressive. Available investment capital has gone mainly into agricultural and health programmes.

Regarding banks whose concern is that of development aid, Ridley (1980) stated the World Bank loans each year to minerals projects have amounted to only 1 per cent ($50 million) of their annual total. The Asian Development Bank and the International American Development Bank between 1971 and 1976 loaned only 0.5 per cent ($8.8 million) of their total per year to non-fuel minerals projects. However, the latter have been more adventurous in recognising the special problems of mining by their proposal to establish a Fund for Energy and Minerals which would ensure equity financing for all forms of minerals development in Latin America as well as offering cover against any kind of commercial and political risk.

As for the UN, development finance for mining was first offered in 1959 with the formation of the Special Fund set up to finance mineral exploration. In the subsequent seventeen years the UN underwrote 123 projects in seventeen countries at a cost of $103,000,000 to that organisation, together with a further $65,000,000 by way of subventions from the host governments involved. As a result of this work sixteen metallic mineral deposits were evaluated as of commercial significance and by 1979 two of these (both copper mines) had come into production in Mexico and Malaysia. Two others were scheduled to begin working in 1981, a manganese operation in Upper Volta and one for tin in Burma. In addition a salt deposit in Morocco and a major limestone deposit in Togo have been brought into production. Carman (1977) has suggested that the total value of all these deposits is in the region of $26,000,000,000 and, on the assumption that they are all developed, the total outlay of funds put together by the UN would appear to be well spent.

If the amount of lead time required (on average 10.5 years) to bring this small number of operations to fruition is indicative of the long time scale now needed to take a deposit from its exploration stage to that of a working deposit, then the amount of money spent by the UN compared with the mining companies could be indicative of its commitment to such explorative operations. By comparison with UN efforts, in the period 1968–71 Wargo (1973) estimated that the average expenditure of 7 mining corporations was $11,600,000 per year and that of a selection of medium size companies around $3,400,000. Reviewing Australian exploration expenditure, excluding hydrocarbons, Emerson (1977) calculated that $130,000,000 was being spent per year, a figure that must have risen significantly in the ensuing period. On this evidence it has to be concluded that the UN contribution in this field is very small. Indeed its total outlay over the period in question represents only 23 per cent more than that spent *in one year in one major mining exploration area*, Australia.

However, a new initiative by the UN was taken in 1975 with the establishment of its Revolving Fund, its purpose being to provide temporary finance for mineral exploration in developing countries. The creation of the Fund was to be from voluntary contributions from industrialised countries with payments from this to be made for the full exploration costs where a project is approved. In this respect it differs from the Special Fund in that the developing country is expected to pay back to the Revolving Fund an amount equal to 2 per cent of the revenues from mineral sales over the first fifteen years of production, but, of course, only *if and when*, a viable deposit is found and brought into production.

Although it is probably too early to judge the effectiveness of this Revolving Fund, its first years were not promising. Only five countries contributed finance (the USA, Belgium, Canada, the Netherlands and Japan) which amounted to a total of $25,700,000. After four years of operating the scheme some fourteen projects had been approved (Radetzki, Zorn 1979), although four could not be proceeded with because of subsequent problems in the host country (Prast 1979). Certainly there seems to be a reluctance amongst less developed countries to commit themselves to use the Revolving Fund because of the commitment to repay exploration costs thus retopping the total finance available (Willox 1980), as well as disappointment amongst UN

| COUNTRIES GROUPED GEOGRAPHICALLY — AFRICA | Copper | Cobalt | Zinc | Lead | Gold | Silver | Chromium | Nickel | Platinum | Tin | Tungsten | Iron | Manganese | Titanium | Molyboenum | Tantalum & Niobium | Antimony | Arsenic | Bismuth | Mercury | Vanadium |
|---|---|---|---|---|---|---|---|---|---|---|---|---|---|---|---|---|---|---|---|---|---|
| Mauritania | ■ | ○ | | ○ | ○ | | ○ | | | ○ | ○ | ■ | ○ | ○ | | | | | | | ○ |
| Mali | ○ | | ○ | ○ | ◧ | | ○ | ○ | | ○ | | ○ | ○ | ○ | ○ | ○ | | | | | |
| Upper Volta | ○ | | ○ | ○ | ◧ | ○ | | ○ | | | | ○ | ○ | ○ | | ○ | | | | | |
| Niger | ○ | ○ | ○ | | ■ | | ○ | ○ | | ■ | ○ | ○ | ○ | ○ | ○ | ○ | | | | | |
| Cape Verdeislands | | | | | | | | | | | | | | | | | | | | | |
| Senegal | ○ | | ○ | ○ | ○ | | | ○ | ○ | ○ | ○ | ○ | | ○ | | ○ | | | | | |
| Gambia | | | | | | | | | | | | ○ | | ◧ | | | | | | | |
| Guinea Bissau | | | | | ○ | | | | | | | ○ | ○ | | | | | | | | |
| Guinea | ○ | ○ | | | ◧ | | ○ | ○ | ○ | | | ■ | ○ | | | ○ | | | | | |
| Sierra Leone | ○ | | ○ | ○ | ◧ | ○ | ○ | | ○ | ○ | ○ | ◧ | ○ | ■ | ○ | ◧ | | ○ | | | |
| Liberia | | | ○ | ◧ | | ○ | | ○ | | | | ■ | ○ | ○ | ○ | ○ | | | | | |
| Ivory Coast | ○ | ○ | ○ | ○ | ◧ | ○ | ○ | ○ | ○ | ○ | ○ | ○ | ○ | ◧ | ○ | ○ | ◧ | | ○ | ○ | ○ |
| Ghana | ○ | ○ | ○ | ○ | ■ | ◧ | ○ | ○ | ○ | ○ | ○ | ■ | ■ | ○ | ○ | ○ | ○ | ○ | | | |
| Togo | ○ | | | | | | ○ | | | | | ○ | ○ | | | | | | | | |
| Benin | ○ | | | | ○ | | ○ | ○ | | | | ○ | ○ | | | | | | | | |
| Nigeria | ○ | | ■ | ■ | ■ | ○ | | | | ■ | ◧ | ■ | | ○ | ◧ | ■ | ○ | | ○ | | |
| Cameroon | ○ | | ○ | ○ | ■ | ○ | | | | ■ | ○ | ○ | ○ | ○ | ○ | ○ | | | | | |
| Equatorial Guinea | | | | | | | | | | | | ○ | | | | | | | | | |
| Gabon | ○ | | ○ | ○ | ■ | ○ | | ○ | | ○ | ○ | ○ | ■ | | | ○ | | | | | ○ |
| Sao Tome Principe | | | | | | | | | | | | | | | | | | | | | |
| Congo | ■ | ○ | ■ | ■ | ■ | | | | | ◧ | | ○ | | ○ | | ◧ | | ○ | | ○ | |
| Chad | ○ | | | ○ | ◧ | | | | | ○ | ◧ | | | | | | | | | | |
| Cent. African Rep. | ○ | ○ | ○ | | ◧ | | ○ | ○ | ○ | ○ | | ○ | ○ | | | | | | | | |
| Sudan | ◧ | | ○ | ◧ | ■ | ○ | ■ | | ○ | ○ | ○ | ◧ | ◧ | ○ | ○ | ○ | | | | | |
| Ethiopia | ◧ | ○ | ○ | ○ | ■ | ○ | ○ | ○ | ◧ | ○ | ○ | ◧ | ○ | ○ | ○ | | | | | | ○ |
| Djibouti | ○ | | | | | | | | | | | | | | | | | | | | |
| Somalia | ○ | | ○ | ○ | ○ | | ○ | ○ | | ■ | | ○ | ○ | ○ | ○ | ◧ | | | | | |
| Kenya | ◧ | ○ | ◧ | ■ | ■ | ◧ | ○ | ○ | | ○ | ○ | ■ | ◧ | ○ | ○ | ◧ | | ○ | | ○ | |
| Uganda | ■ | ■ | ◧ | ◧ | ◧ | ◧ | ○ | ○ | ○ | ◧ | ■ | ○ | ○ | ○ | ■ | ◧ | | | ◧ | | |
| Rwanda | ○ | ○ | ○ | ○ | ◧ | | ○ | ○ | | ■ | ◧ | ○ | ○ | | | ■ | | | ◧ | | |
| Burundi | ○ | ○ | | ○ | ◧ | | ○ | ○ | | ■ | ◧ | ○ | | | | ◧ | ○ | | | | |
| Tanzania | ◧ | ○ | ◧ | ◧ | ■ | ◧ | ○ | ○ | | ■ | ◧ | ○ | ○ | ◧ | ○ | ◧ | ○ | ○ | ○ | ○ | ○ |
| Malawi | ○ | ○ | ○ | ○ | ○ | | ○ | ○ | ○ | | | ○ | ○ | ○ | ○ | ○ | | | | | |
| Zaire | ■ | ■ | ■ | ■ | ■ | ■ | | ◧ | ◧ | ■ | ■ | | ■ | | ○ | ■ | | ○ | | ○ | ○ |
| Zambia | ■ | ■ | ■ | ■ | ◧ | ◧ | ○ | ○ | ○ | ■ | ○ | ◧ | ◧ | | ○ | ◧ | | ◧ | | ■ | |
| Botswana | ■ | ■ | ○ | ○ | ◧ | ◧ | ○ | ■ | ○ | | ○ | ○ | ◧ | | | | ○ | ○ | | | |
| Swaziland | ○ | ○ | ○ | ○ | ◧ | ◧ | ○ | ○ | ○ | ■ | ◧ | ■ | ○ | | ○ | ◧ | | | ○ | | |
| Lesotho | | | | | | | | | | | | ○ | | | | | | | | | |
| Madagascar | ◧ | ○ | ○ | ◧ | ◧ | ○ | ■ | ○ | | ○ | | ○ | ○ | ◧ | ○ | ◧ | | | | | |
| Comoros | | | | | | | | | | | | | | | | | | | | | |
| Mauritius | | | | | ○ | | | | | | | ○ | | | | | | | | | |
| Seychelles | | | | | | | | | | | | | | ○ | | | | | | | |

*Table 4.1* Metallic mineral resources of some African countries: ■ present production: ◧ previous production: ○ resources of unknown potential.

*Source*: Hunting Geology and Geophysics Ltd.

officials over the reluctance of developed countries to contribute capital in the first instance.

It is significant that the country that has contributed over 70 per cent of the initial funding is Japan, the only country with a consistent long term policy towards securing its mineral supplies. This is evident from the many joint ventures undertaken by Japanese companies with overseas countries where the risk capital invested by them has always been underwritten by their government (see Chapters 5, 6 and 7 on individual minerals). Whilst at first glance it may seem noteworthy that the UK and West Germany have not participated in the Revolving Fund, as members of the EEC they have been part of a collective approach to the support of mining ventures in the less developed countries. This was achieved by the EEC through the Second Lomé Convention, an agreement designed to provide grants and loans for certain less developed countries' developments in the years 1980–85. Signed in 1979 the Convention recognises that it is in the mutual interests of the Community (as minerals importers) whose rate of investment in the mid-1970s was probably only around 20 per cent of that required to satisfy future needs (European Group of Mining Companies 1976) and of fifty-eight less developed countries in Africa, the Caribbean and the Pacific (as minerals exporters) to encourage new mining activity. Thus, apart from the provision of technical mining assistance from the EEC to those less developed countries, it has made $226,000,000 available for the financing of minerals and energy projects of interest to both parties and to encourage the investment of private capital, it has set out conditions regarding the shape and form of agreements between the Community and its member countries on the one hand and the signatory less developed countries on the other; and finally, it has provided an insurance scheme (backed by a sum of $372,000,000) covering any catastrophic fall in the price of the main minerals likely to be produced by any of the less developed countries involved, such as copper, cobalt, phosphates, manganese, bauxite, tin and iron ore. The development potential of the Second Lomé Agreement is amply demonstrated in Table 4.1 which shows for the African group of countries alone, not only what is being extracted in the early 1980s but where future opportunities lie.

This comprehensive package of technical aid, investment, insurance against low prices and the laying down of terms and mining conditions acceptable to both the host country and the potential mine operator seems most likely to succeed. On the one hand it should satisfy the less developed country which will not wish to cede its resources prematurely and recklessly, and on the other, principles and general terms agreed and backed by all the parties to Lomé II should provide the necessary assurance that the mining company will not be subjected to confiscatory conditions once the ore body has been discovered and mining commences. The arrangements suggested under Lomé II certainly offer a model for suppliers and consumers which could create a greater degree of stability of minerals supply and demand if applied more widely in future.

## CHAPTER 5

# Key industrial minerals – non-metalliferous minerals

A mineral classification such as that of Jones and Darkenwald (1965) discussed in Chapter 3, would underpin the physical and chemical characteristics of the minerals in question, as well as consider their end uses. In this and the next two chapters, in which the purpose is to examine each of the major industrial minerals, such an approach would have brought iron and steel into association with those other metals used in the preparation of special steels before considering the older major metals, the light metals and the precious metals. A large range of minerals used in the chemicals industry could then have been considered followed by the construction minerals, the refractory minerals, minerals used in the electronics industry, the nuclear minerals and finally the carbon and hydrocarbon or fuel minerals. An approach such as this characterises the survey published each year as part of the *Mining Annual Review*.

However, those with a greater concern for a spatial dimension within a minerals classification require an approach which is somewhat different. For this reason the framework has been adopted which was originally used in *The Mineral Resources of Britain* (1975). Amongst other things it introduces the important distinction of location and draws upon the concept of place value. Five main categories emerge from such a classification: the ubiquitous non-metalliferous minerals; the localised non-metalliferous minerals; the ferrous metals; the non-ferrous metals and the carbon and hydrocarbon minerals. Within this approach some attempt has nevertheless been made to acknowledge usage and a number of sub-groupings reflect this. Thus following a discussion in this chapter of the ubiquitous non-metallic minerals (section 5.1) which are of prime importance to the construction industry (that is sand and gravel, stone, cement, clays and shales), consideration is given to the localised non-metalliferous minerals (Section 5.2) associated with that end use, gypsum and asbestos. A second group of localised non-metalliferous minerals (phosphate, potash, sulphur, salt, fluorspar, barytes and boron) are taken together because of their primary concern in the manufacture of chemicals. Even here a further distinction is made between the first two minerals and the rest since the former are closely associated with fertiliser production. Finally, under the general heading of localised non-metalliferous minerals, china and ball clays are grouped together because of an association of end use.

In Chapter 6 on metalliferous minerals, discussion of the ferrous metallics is followed by an account of those non-ferrous metallics associated with them primarily in the production of a wide range of special steels (that is manganese, chromite, nickel, cobalt, molybdenum, tungsten and vanadium). A series of metallics which are characterised by their lightness are then discussed (aluminium, magnesium, titanium). The major traditional metals (copper, tin, lead and zinc) come next followed by the precious metals (gold, silver and platinum minerals); a group of minerals associated with the chemicals industry (antimony, rare earths and lithium); the electronic metallics (mercury, cadmium, indium, rhenium, selenium and tellurium); and the nuclear metallic, uranium. Finally, this part of the book ends with a grouping of minerals (Ch. 7) which has generic and end use communality and includes coal, oil and natural gas.

The discussion of each mineral does not necessarily conform to a fixed pattern though it does set out to cover roughly similar ground. There is an attempt to explain the geological circumstances in which each mineral is found, the methods by which it is worked, and the nature of the end use of the mineral. Although the first two aspects of this discussion may not seem immediately relevant to a consideration of resource management questions, factors regarding the geological circumstances of a mineral and thereby the technique by which it may be worked in turn affect the economics of the mining operation and ultimately whether a given mineral may be regarded as part of total reserves. Also of central importance are the spatial patterns of supply and demand for a mineral and their dynamics, the relationships in political and economic terms between the countries involved, and the changing location of the intermediate processing of a mineral once it is won from the ground. In some instances the major environmental issues which emerge as a result of production are also considered and here again these may be seen in some instances to be related to geology and extractive methods. Although the longer term viability of mineral supplies is discussed in Chapter 11, current levels of availability and those for the immediate future, covering the period up to the end of the twentieth century, are a matter for discussion at the end of the consideration of a mineral or group of minerals.

The likely resource totals and the reserves which form part of those totals are based on United States Bureau of Mines (USBM) calculations as are those for increases in levels of demand for each mineral expressed as a percentage annual increase. The former task of calculation is not easy, involving not only the collation of available national data, but in some cases the validation of suggested resource and reserve figures and the estimation of both based on broad geological data where the political regimes of the countries in question do not allow further investigation. However, the latter task, with its evident forecasting component is even more problematic. Here the work is based on probabilistic and multiple contingency analysis of technological and economic assumptions made for the major end uses for each mineral and each country. The statistical projections which are quoted are thus established by correlating the utilisation of each mineral with economic indicators such as gross national product for each consuming country (or country that is likely to be consuming in the future for part or all of the period in question) and their population levels. The assumptions and exogenous variables used to establish the statistical projections vary according to the mineral and to its end use in each country.

To illustrate the problems of prognostic calculation, for sand and gravel demand is intimately linked to its main consumer, the construction industry, which is in turn closely geared to the health of national economies. However, economic recessions such as those of the mid and late 1970s precipitated by the oil crises of 1973/74 and 1978/79 are almost impossible to predict. Population levels are also a highly significant control on levels of raw materials consumption by the construction industry and here again official forecasts have often been wide of the mark. Of the other imponderables that affect the accuracy of attempts to predict consumption are: (1) changes in the levels of investment in specific private and public sectors of the economy; (2) alterations in policy at government level regarding funding priorities; (3) changes in construction technology for highways or buildings; and (4) the degree to which other aggregate materials may be substituted or alternative artificial materials which are more easily handled or have better technical specifications can replace sand and gravel. The question of substitution is plainly one that influences the likely future levels of demand for many of the minerals in very different ways and according to the end use to which the material is put. For example, minerals such as manganese, chromite, nickel, cobalt, molybdenum, tungsten and vanadium are commonly used as additives in the manufacture of different types of steel. Although substitution of one for another is largely possible, given the specific end use of the steel, the substitution is usually only made at additional cost.

Another form of substitution involves not the replacement of one mineral by another in a given product, but the substitution of a totally different product to serve the same end. This again lowers demand for the initial mineral. Thus in the case of the recording of visual images, until the late 1970s film stock made substantial demands on silver which forms a vital part of its light-sensitive make up. However, the development of the simple video camera which uses tape coated with iron or chrome oxide as the recording medium is shifting demand away from silver. The speed with which the transfer will occur and its ultimate extent is difficult to predict.

Other prediction problems also arise from the differences between developed and less developed countries. Should it be assumed for instance, in forecasts of demand that over the short to medium term the latter will begin to emulate the former in so far as per capita demand for steel is concerned, or will this be of much longer term significance as new large-scale capital projects are undertaken in the less developed countries? Is it more likely in the short to medium term that demand for such products as kaolin as a paper filler will take off in less developed countries now that for nations such as the USA, with its high levels of literacy, the market has become saturated? Whilst the complexities of forecasting the demand for minerals cannot be under-estimated in general terms, the problems peculiar to specific minerals will be mentioned in the text that follows.

## 5.1 Ubiquitous non-metalliferous minerals

### 5.1.1 Sand and gravel; Stone

*Sand and gravel*
These originated as a result of the disintegrating action of weathering on a wide variety of different parent rocks. They are granular materials consisting mainly of silica but contain other minerals such as magnetite, mica and feldspar in varying amounts, together with iron oxides and some heavy metals. The sand is usually in the size range 0.0625 to 2 mm, whilst the gravel is made up of particles larger than 4 mm but less than 64 mm in diameter.

Sand and gravel is classified according to its origin or its method of deposition. Deposits of material that have been carried by rivers and streams are referred to as fluvial. At any one site such deposits of sand and gravel exhibit a limited size gradation but the distribution of size ranges varies greatly depending on whether the stream had been fast flowing, meandering, narrow or shallow along that stretch. Glacial deposits were distributed by the massive ice sheets which covered large areas of North America, northern Europe and Asia. These consist of particles of all shapes and sizes, with little grading or sorting. Lake and marine deposits consist of hard material which is worn and rounded and well segregated according to size. Deposits of materials which are found on top of a parent rock body, usually consisting of a well mixed variety of particle sizes, are referred to as residual.

Sand and gravel deposits are widespread throughout the world and tend to be worked as close as possible to centres of demand, primarily because of their low unit value (Fig. 5.1). Urban development has greatly increased the need for sand and gravel to support the requirements of the local construction industry. However, local shortages can arise since urbanisation frequently overtakes land from which the resource can be extracted. Moreover, even where resources are in good supply, the proportions of the various size fractions are seldom optimal for all the needs of the area. Apart from the 'permanent' pits peripheral to urban zones, more transitory exploitation tends to occur where highways, dams and other major public works are

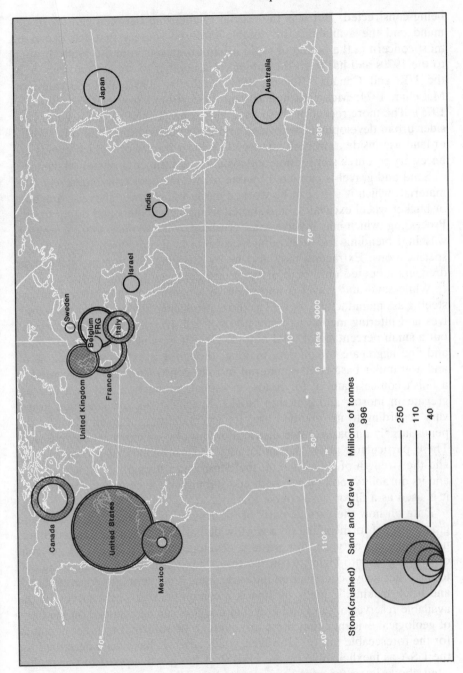

*Fig. 5.1* Sand and gravel, stone: output by country with chief centres of production, 1980. *Source: Mining Annual Review, Minerals Year Book*

being constructed. But it is the crucial relationship between centres of demand and the availability of suitable localised sources that has exercised most concern in the period of rapid urban expansion from the 1960s through to the 1970s and has been the subject of a number of studies particularly in the UK and Canada (Beaver, 1968; McLellan 1967, 1970; Bryant and McLellan 1974; McLellan and Bryant 1975; Blunden et al 1974; Blunden 1975). The more recent of these have also concentrated on the need to consider urban development with sand and gravel exploitation and other forms of land use inside a properly managed regional framework which avoids unnecessary resource sterilisation and avoids long-term environmental damage.

Sand and gravel is extracted by one of two methods from open pits. The material, which is soft, can be worked easily using power shovels, draglines or bucket wheel excavators and shifted using high loaders and dump trucks. Processing, which may involve some crushing, will also necessitate screening, washing, blending and stockpiling according to the required standards and specifications. Extraction may also be by dredging. Suction or bucket type dredgers mounted on a boat or barge are deployed for this purpose.

While sand and gravel is important to a number of industries (iron and steel, glass manufacture, as well as those producing silicate chemicals, abrasives and filtering media) its main use is in construction which consumes all but a small percentage of total output. Here it is valued primarily as coarse and fine aggregate in concrete making, though it is extensively used as fill and a granular base course material in road construction. It is also used as a finish course material for roads, in asphalt production and as a fine aggregate in mortar and concrete blocks. Specifications for sand and gravel vary according to its intended use. Its acceptability in a given situation depends largely on particle size distribution but also on chemical composition. These particularly affect the uniformity and workability of a concrete mix and the strength of the concrete, the density and strength of an asphalt mix and its durability, and the strength and stability of the compacted mass when it is used as a fill or base course material.

Data on how the various parts of the construction industry use sand and gravel are not forthcoming on a worldwide basis, although the United States figures are not untypical for western developed nations. There 63 per cent of output goes into highway construction, 25 per cent in general building and to the other heavy construction industries, and 7 per cent into concrete block and brick making. Equally lacking is a comprehensive world survey of the available reserves of sand and gravel although it can be assumed on the basis of geological evidence that these are likely to be sufficient to meet demands for the foreseeable future. However, whilst reserves have been described for the USA as inexhaustible by the USBM, it has given support to the studies cited above in so far as it has suggested that the geographical distribution and quality of land-based deposits often do not match market patterns and requirements. One answer in the US context has been to develop ocean mining when this can be adjacent to areas of shortfall. Other countries similarly placed have also begun dredging the shallow waters of their continental shelves; these include Denmark, Sweden, Japan, Thailand, Hong Kong, and the UK which has by far the largest and most advanced off-shore operations.

In the UK, the proximity of marine supplies particularly to London and the south-east has been of considerable importance. Here demand has continued to escalate whilst problems of obtaining consent to work land-based resources have become acute and extremely expensive if they do become available. Not surprisingly sea-won sand and gravel has trebled in output in the twenty years up to the mid 1970s the major share coming from the coastal waters around the south-east of England. By the end of the period in question, supplies obtained in this way represented 12 per cent of total UK production and appeared to be increasing at around 6 per cent per annum.

Whilst sand and gravel is the world's principal construction material, alternatives which offer it the strongest competition are the various forms of crushed stone. These have essentially the same end uses though limestone is the most common substitute as a construction aggregate. In the making of concrete, expanded clay, shale, slate, vermiculite, perlite, natural pumice, blast furnace slag, pulverised fuel ash, boiler house clinker, crushed brick and colliery slag can all be used as alternatives to sand and gravel.

The production costs of sand and gravel vary widely depending on geographic location, composition of deposits and number and type of products produced. However, it is generally a low-cost material, a situation determined not only by its relative abundance and ubiquity but also the simplicity of its extraction and processing. In the two decades preceding the mid 1970s, the unit value of sand and gravel generally declined in real terms with increased labour costs and rising land values, offset by increased mechanisation and automation. Further improvements in operating efficiency are now unlikely and with more closely defined specifications, more steeply rising labour and property costs, the need to develop less favourable sites on land or deeper deposits at sea, and the enforced compliance with more stringent anti-pollution and rehabilitation regulations, sand and gravel prices are likely to show a slow but steady increase.

Estimates of future demands for sand and gravel have been made by the USBM which suggest an annual growth rate of 3.5 per cent to the end of the century. Such prognostications must, however, be viewed with great caution since requirements for sand and gravel are closely tied to activity in the main consumer, the construction industry, which is in turn, as the introduction to this chapter made clear, closely geared to the health of national economies.

## Stone

Rock material which occurs naturally and is quarried or mined for industrial usage without change in its chemical state and with its physical characteristics transformed only by shaping or sizing is commercially termed 'stone'. Two

69

forms of such stone may be distinguished. Building stones consist of material cut and shaped for use as building blocks, slabs or panels, taken from deposits of limestone, granite, marble, sandstone, slate or basalt. More important are the broken, irregular, screened and sized rock materials that make up the crushed stone category and are primarily used for aggregates. It is these which, in most countries, represent well over 80 per cent by weight of all stone produced. Because of the diversity of materials available as crushed stone, specifications are numerous with variations in the physical and chemical characteristics within each type and a large number of different end uses. Nevertheless, limestone, accounting for about 70 per cent of output, dominates the crushed stone industry.

Stone suitable for crushing is widespread throughout the world (Fig. 5.1). The USA leads in terms of production at over 1,000 million tonnes per year and with the possible exception of the USSR and the People's Republic of China, whose outputs are unknown but believed to be large, no other country produces more than 200 million tonnes a year. Other leading producers are the German Federal Republic, France, Japan, the UK and Australia. Building stone is equally ubiquitous with nearly every country represented as a producer. Nevertheless, output is dominated by three nations. Italy has high quality resources of marble, travertine, tufa, limestone, serpentine and volcanic tuff. Limestone, marble, granite and serpentine are extensively worked in Portugal. In France, granite, limestone and marble are important.

The production methods used in the extraction of building stone range from the primitive and inefficient to those which are modern and technically sophisticated. The raw material is rarely blasted out of the quarry face. Rather, blocks are cut then sawed or split into smaller blocks for ease of handling and transport and taken to the processing plant for final cutting and finishing operations. Decoration is carried out by skilled craftsmen, sometimes using pneumatic cutting tools or sand blasting.

While some crushed stone is worked underground, this material is won largely using open-pit techniques. Unlike sand and gravel, the material has to be drilled and blasted from the rock face. In order to minimise blasting in environmentally sensitive areas, further breaking down of the rock is achieved using drop hammers. After primary and secondary crushing at the processing plant the material is passed through inclined vibrating screens before being stock-piled according to size.

Crushed stone can be used in a multiplicity of roles. Chief amongst these is the use of limestone for cement, but this material is also of value in the production of iron and steel (as a metallurgical flux), and for agricultural purposes. However, over four-fifths of crushed stone is used for construction purposes with highways being the chief consumer, followed by buildings and concrete. All major types of crushed stone are used in construction but again the major portion is limestone. Indeed, of the crushed limestone produced in the USA, 73 per cent is used in the construction industry, 13 per cent for cement manufacture, 4 per cent for lime manufacture and 4 per cent for agricultural purposes.

Although building stone is but a small fraction of the total amount of stone worked, about three-quarters of it is again used by the construction industry, with exterior facing panels for buildings taking the largest share. Apart from the use of building stone for curbing, flagging and roofing, it is clear that in the last forty years its primary role has changed from one in which it was used for its structural qualities to one where it is valued mainly for its decorative or aesthetic qualities.

Stone resources as a whole may be considered on a world basis as inexhaustible. The actual amount available is so vast as to make measurement impossible. It is possible, of course, to envisage a shortage of one specific stone or another, either nationally or regionally. However, the local shortages that occasionally exist are not usually due to a lack of stone but to urban encroachment on resources and to zoning or planning regulations (see Ch. 8) that may prevent the development of quarries, or limit the availability of certain types of material, usually limestone, unless these are won by underground methods. Such problems are likely to become more acute in the future.

In terms of alternative materials, as far as the construction industry is concerned, sand and gravel predominates. It is available usually in large quantities and at prices which are highly competitive. Other substitutes are largely the same as those for sand and gravel. However, there seem to be no raw materials that seriously compete for the making of lime and cement, with the notable exception of chalk which is really an extremely friable limestone. Substitute raw materials are available for most of the building stones and these have made substantial inroads into their markets, particularly because of their low costs and ease of use in construction. Slate, to cite one example, finds itself largely ousted for roofing by clay, cement and asbestos tiles. Now building stone producers have to compete almost entirely on the decorative and aesthetic qualities of their wares, especially in the cladding of prestige buildings. Certainly in cost terms alone, their materials, compared with the alternatives, are very uncompetitive. This is mainly the result of the labour-intensive methods of production. While new technology is being introduced into the preliminary stages of working, it is expected that the reduction in unit material costs will do little more than offset rising labour charges in the immediate future.

In contrast, crushed stone prices are extremely low. Although they approach those of their main competitor, none of the chief aggregate materials have the production costs of sand and gravel. The nearest and chief rival, limestone, is more expensive to work since it requires blasting from the quarry face, greater processing, and is harder on production plant.

Other forms of crushed rock are progressively even more expensive to produce. Thus at a centre of demand for crushed rock aggregate materials where no distinction in terms of end use can be made between rock types and where all forms are equally available, limestone will always be preferred to sandstone, and sandstone to igneous rock.

Although in real terms crushed rock production costs have been relatively stable, and in some instances have actually declined during the twenty years up to the mid-1970s due to mechanisation and the introduction of automated plant, the increasing costs of land acquisition, as well as environmental controls during production and rehabilitation schemes for the worked out land afterwards, are bound to have an increasing impact. Underground extraction methods can eliminate most of the environmental and rehabilitation problems because of their minimal surface impact, lower land costs (particularly in urban areas where surface prices are at a premium) and avoidance of the costs and the difficulties attached to overburden stripping and storage. However, underground extraction can raise extraction costs per unit of saleable material by 50 per cent compared with open pit techniques. Nevertheless, such methods can be made to pay where the working is adjacent to or even *under* centres of demand since transport costs can be drastically reduced. The operation itself can also create potentially valuable storage areas. When such space is at a premium it can provide revenues in excess of that of the stone extracted. By 1976, in the USA, where the underground production of crushed stone is furthest advanced, some 5 per cent of total output was gained in this way.

Regarding the future production of crushed stone, USBM forecasts on a world basis have been made of an average 3.5 to 5 per cent growth rate per annum up to the end of the century. For individual countries, variations in output will be considerable, but with the exception of limestone produced for fluxing and agricultural purposes, demand for crushed stone will largely be tied to construction activity, as indeed is that for building stone.

As for building stone, demand up to the end of the century may rise between 1 and 2 per cent per annum. High forecasts are certainly dependent on production economies keeping prices down whilst the lower demand figure presupposes further inroads being made in the market by substitute materials.

### 5.1.2 Cement: Common clays and shales

*Cement*

The cements referred to here are primarily of the common calcareous hydraulic form. They are made from limestone and have the property of setting or hardening under water. The majority of these cements are of the Portland type though others are important for their specialised applications. Portland cements are produced by burning, usually in a rotary kiln, a measured quantity of the chief raw material, limestone, together with silica, alumina and iron oxide. Alluvial and residual clays and sedimentary deposits of shale and schist are common sources of alumina and silica. (Slate, andesite and granite, some igneous rocks such as pumice and tuff are used to a lesser extent, as is blast furnace slag.) Iron ore is the predominant source of iron oxide. The discharge from the kiln, known as clinker, varies in particle size from fine sand-like grains to that of walnuts. It is a chemically complex mixture of calcium silicates and aluminates. This is mixed with 4 to 5 per cent gypsum and ground to a fine powder to form cement.

There are two methods of making cement. The wet process involves the addition of water to produce a slurry ıring grinding. In the dry process, the basic raw materials are dried and pulverised to a powder. Older plants sometimes have vertical or shaft kilns but these have a low annual capacity of about 54,400 tonnes compared with 1.13 millions for the larger rotaries.

The basic inputs to the process of cement manufacture are bulky and of low value yet in the course of making the end product these suffer substantial weight loss. The weight ratio of raw materials to the Portland cement product is 20:11. Processing plants must therefore be located close to their raw materials, but within reasonable range of markets. Fortunately, the relative ubiquity of limestone and clay does not make this too problematic. Moreover, cement is considerably more expensive than other materials so far discussed in this section. Indeed its unit cost at the point of production is some twenty times that of sand and gravel. Not surprisingly the market radii for Portland cement are quite large by comparison and may be considered regional rather than local. In the USA they average about 300 kms although they may be extended by as much as five times with cheaper water transport. Production sites adjacent to port facilities have developed export markets for Portland cement. Certain speciality cements are regularly carried great distances since transport costs do not represent as high a proportion of the landed price and the quantities involved are generally much smaller than for Portland cement. But whilst the speciality cements such as white cement, oilwell cement and expansive and waterproof cement have individually important functions, they make up less than 1 per cent of the total market leaving Portland cement to account for most of the rest.

Portland cement is consumed almost in its entirety in construction where it is combined with water and aggregates to make concrete. As the most widely used and adaptable building material, it can be poured on site for large construction projects such as buildings or dams, used in the form of heavy pre-stressed columns or beams, or in the form of delicate precast panels or pipes. A small percentage of cement is also used in the building trade mixed with water and fine aggregate to form mortar. About 1,700 clinker producing cement plants can be identified in 135 countries. Annual output was running at 938 million tonnes in 1980. The USSR was the largest producer with about 15 per cent of the total output followed by Japan at 10 per cent and the USA at 9 per cent. The European Cement Association comprising nineteen member countries produced 32 per cent of output; Asia (except Japan and the USSR), 22 per cent; North and South America (ex-

73

cept the USA) 8 per cent; Africa 3 per cent; and Australia less than 1 per cent. The sixteen largest cement producing countries accounted for 73 per cent of total output (Fig. 5.2).

In a world market in which some cement and clinker shortages were evident in the USA, Near East, North and West Africa, South America, the Far East and to a lesser extent Eastern Europe, less than 4 per cent of the total output was internationally traded. Of this amount just under half consisted of exports by the European Cement Association of which two-thirds was between member countries.

The shortfalls in production arise from two principal factors. One of these connected with environmental pollution will be discussed later in Chapter 8. The other concerns the single most important cost factor in the production of cement, fuel. Energy represents 40 per cent of direct production expenses. When the oil crises of the 1970s brought about massive increases in fuel bills some less developed countries were faced with very considerable balance of payments problems.

A number curtailed or closed down their clinker production facilities making do with limited imports of clinker which they subsequently ground into cement. How long this situation will pertain remains to be seen. In the longer term both in less developed countries and elsewhere the trend will be towards the dry process kilns in which fuel savings of between 40 and 50 per cent over other techniques can be attained.

Apart from fuel prices those for labour have been of greater impact on unit costs in recent years although they have been largely offset by larger units of production and automated centralised process controls. These can frequently be serviced by as few as two men per shift.

Future increases in the rates of world cement consumption have been estimated at 2.7 per cent per year by the USBM up to the end of this century. Since cement production is largely tied to the fortunes of the construction industry, it is the health of this industry that will ultimately impact most on future demand levels though for less developed countries the question of fuel availability will be pertinent.

Even though there are no substitutes for cement in the production of concrete, there are, of course, alternatives to concrete in the construction industry. However, with such materials as wood becoming scarcer and the possibility of further inroads being made into the building stone market, it seems unlikely that the demand for cement will slacken.

The raw materials from which cement is made appear in abundance on a world-wide basis. Future demand can be met although in specific areas the sterilisation of resources as a result of urban development and the imposition of embargoes on the winning of limestone adjacent to homes or in areas of natural beauty may have their effect unless underground working can be developed as an alternative.

Environmental considerations, however, have had an even more profound impact on plants currently producing cement and their location.

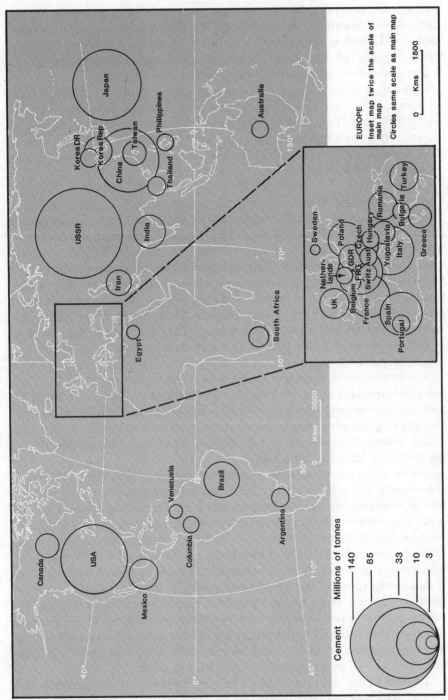

*Fig. 5.2*  Cement: output by country with chief centres of production, 1980.
*Source: Mining Annual Review, Minerals Year Book.*

Whilst some general indication of these on a local basis was given in the discussion of limestone, the reaction in the USA by the producers to standards for air and water quality set in the 1970s by the Environmental Protection Agency (see Ch. 8) has been marked and has led them to close plants in a considerable number of instances, rather than comply.

Grinding mills have continued to operate using imports. Thus while of the total imports in 1961, only 10 per cent was clinker, this had risen to 15 per cent in 1970 and 41 per cent in 1973 (by which time the regulations had really begun to bite) and had eased to 33 per cent by 1975. Imports were mainly from Canada (60%) nearly all of the rest coming from members of the European Cement Association.

### Common clays and Shales

These are the product of the mechanical or chemical weathering of parent materials such as feldspar and mica. The products from these reactions consist of mixtures of particles of various sizes and different physical, chemical and mineralogical properties. They can be classified into one of three groups based on their chemistry and structure; the kaolinite group, the montmorillonite group and the illite group. It is only the last of these (the common clays and shales) which are of concern here. These materials are indigenous to most countries and plentiful within each. They are, for example, worked in forty-seven of the fifty United States of America and in all of the UK economic planning regions.

The prime use of clays and shales is in the manufacture of bricks. They are also important in the making of other building products such as pipes, tiles and light-weight aggregates, whilst some go into the making of cement. In each of these processes value is added to what is otherwise a very low-priced commodity. Sources of raw material must therefore be directly associated with manufacturing plant.

Whilst the increase in value is considerably more than that achieved in the washing and grading of aggregates, the finished product usually remains one of relatively low value. Consequently plant for making bricks, tiles and pipes has traditionally been market-orientated and well-dispersed. Exceptions to this can occur. In the case of the brick manufacturers located on the Oxford Clays in the UK, the market areas have expanded well beyond those which could normally be expected and have become almost national. This is largely the result of the exceptional quality of the clays which do not require weathering, offer the right degree of natural plasticity for shaping into bricks, have very few impurities and a carbonaceous content of 5–7 per cent which reduces fuel bills in the processing. These unusual natural advantages are also combined with a central location with respect to major UK conurbations as well as good road and rail links to these.

The extraction and processing techniques deployed for clay and shale are simple. The raw material is worked by open-pit methods. Blasting is not required, the material being won by draglines and power shovels or shale planes and scraper-loaders. Where appropriate either screening, crushing and drying are used, or the material is extruded and cut into bricks, tiles or pipes before firing in kilns to produce the finished product. Kilns can be fired with electricity, natural gas or coal.

Whilst overall clay and shale reserves are enormous and their occurrence ubiquitous, some distinction should be made concerning availability according to end use.

For common clays and shales, although they are generally usable for brick-making, in terms of local market requirements, they may not meet the specifications of the user either for colour or strength. Likewise all types of clay or shale are not suitable for use in cement manufacture and for making light-weight aggregates relatively few deposits are appropriate. These problems, however, can generally be identified as of local or regional rather than national significance. Thus most countries meet their own requirements for clay and shale and world supply and demand for these materials is in balance. According to the *Mining Annual Review* (1982) total world consumption of common clays and shales used in construction was over 500 million tonnes in 1980, with a predicted annual growth rate of over 2 per cent.

The accuracy of such forecasts must inevitably be tied in with those of the construction industry and the influences bearing on this have already been discussed. Other factors which will affect future demand for clays and shales can be isolated according to their end use. Products used in the construction industry, for example, may suffer competition from alternative materials. Such a situation could be intensified if the cost of brick laying continues to rise sharply. On the other hand, factory-built brick panels could largely offset such a situation. The clay used in the making of brick could find itself under pressure from other suitable materials usually considered wastes (that is china clay quartz and pulverised fuel ash) if pressures to diminish the tipping of these on environmental grounds persist (see Ch. 9). The utilization of clay or shale in the manufacture of cement will also be dependent on how well concrete competes with steel as a structural load-bearing material and the effectiveness of attempts to promote cement facings for buildings. Asphalt could also replace the use of concrete for paving purposes. Again the clay itself could be replaced in cement making by pulverised fuel ash.

The importance of light-weight aggregates where weight reduction is necessary (for example bridge decks, high-rise buildings and large unsupported roof spans) appears to be surging ahead strongly. It is possible that such materials will also be in demand for highway construction where they offer better skid-resistant characteristics than conventional aggregates, particularly on asphalt roads.

Increases in the size of individual clay and shale workings have been accompanied in recent years by greater mechanisation and the achievement of increased economies of scale. But whilst these are important in terms of cost reductions per unit of raw material won, especially offsetting rising labour costs, problems of rehabilitating worked-out sites have become all the greater. In one UK county, Bedfordshire, on the Oxfordshire Clays, there were in 1970 well over 70 million cubic metres of worked-out pits. Whilst efforts have been made to fill such excavations with pulverised fuel ash or domestic refuse, the problem remains largely unsolved. In the UK and North America, companies allowed to work clay and shale pits are now at least required to ensure that waste materials are graded, the excavations recontoured and grass and trees planted. However, the more general environmental impact of mining operations is addressed in Chapter 8.

## 5.2 Localized non-metalliferous minerals

### 5.2.1 Gypsum; asbestos

*Gypsum*
This is the naturally occurring mineral hydrous calcium sulphate. When it is calcined to temperatures of from 100 °C to 120 °C, it loses over 50 per cent of the water of crystallisation and is converted to plaster of paris. This, when ground and mixed with water and allowed to dry, forms a hard plaster product. Anhydrous calcium sulphate, a mineral with no water of crystallisation, is called anhydrite and is commonly associated geologically with gypsum. Both forms were widely laid down around the world and though deposits may be found in any geological era, they are commonest in the Permian and Triassic. Deposits frequently occur as large lenticular stratified bodies which were formed by the evaporation of sea water.

Gypsum is also recovered in some countries as a by-product of chemical plants (usually from the manufacture of phosphoric acid from phosphate rock) and is ultimately utilised in the same way as the naturally occurring mineral. Deposits of the raw material found near the surface are developed by removing the over-burden and then deploying the normal open-pit techniques used in the working of a soft rock. Underground bodies of gypsum are exploited by the normal room and pillar method. Modern extractive enterprises winning gypsum are capital as opposed to labour intensive and designed to maximise economies of scale.

Subsequent processing depends on end use. Gypsum required for the manufacture of plaster and plaster products is crushed and pulverised and then calcined in rotary kilns where 75 per cent of the water of crystallisation is driven off. Plaster or wall board, the primary end product, is of considerable importance to the building industry. Its use has grown faster than that of any other building material. In the USA over 90 per cent of the output of calcined gypsum is used in this way. It is made by introducing calcined gypsum paste between two sheets of paper with thickness determined by passing this 'sandwich' between rollers. When the plaster has set, the resultant wall board is cut into the desired sizes. The small percentage of calcined plaster used for coating ceilings and walls usually requires the introduction of a retarder before use, frequently a glue-type material. Most of the crude gypsum (around a third of the total production of gypsum in the USA) finds its way into the cement industry where it is used itself as a retarder. The other uses of gypsum outside the building industry (for example, agriculture) are of minor significance making little impact on overall demand.

It follows from comments made regarding the relative ubiquity of gypsum (though it is far less common than the minerals so far discussed), as well as its uses, that production and consumption are principally associated with urban development. Where national consumption exceeds production, then importation occurs. Thus for the largest producer and consumer, the USA, it extracts about 19 per cent of world output whilst consuming 28 per cent. Other leading producers are Canada (12%), the German Federal Republic (11%), France (9%), Iran and USSR (8% each) and Spain (7%), although many of the countries of Asia, South America and Oceania have significant levels of production (Fig. 5.3). Where imports of the raw material occur, supplies are sought close at hand because of the fairly low value and substantial bulk of the unprocessed product. Thus Canada supplies all of the USA's import needs. Also, because of the value added by the manufacture of gypsum into plaster, plaster board and so on, plant undertaking such activity tends to be located at the raw material source.

World reserves of gypsum, estimated at 2,400 million tonnes by the USBM, can be regarded as more than adequate for the foreseeable future. The main Canadian production areas (located chiefly in Nova Scotia, but also New Brunswick, Ontario, Manitoba, British Columbia and Newfoundland) have around 410 million tonnes whilst the main USA sources (the Great Lakes area, Texas, Utah, California) have about 700 million tonnes. With the gypsum beds of the Paris Basin extending over 8,000 km$^2$ and to a depth of up to 55 metres the resource base there may be very considerable.

World output which amounted to over 77 million tonnes in 1980 is expected to rise by a percentage annual growth rate of around 2.3 according to the USBM. Forecasts such as this are largely dependent on what happens in the construction industry and the factors governing future developments there have been explored for other building raw materials. However, trends within that industry will also be important.

The increasing part played by prefabricated construction items, especially plaster board, cannot be underestimated in the light of rising labour costs and in spite of the fact that here alone, of all the uses of gypsum by the building industry, there are substitute materials (for example, concrete, brick, fibre board). On the other hand plaster coating for walls and ceilings seems likely to decline further since its application is labour intensive. The use of raw gypsum as a cement retarder will be closely tied to the levels of production of cement.

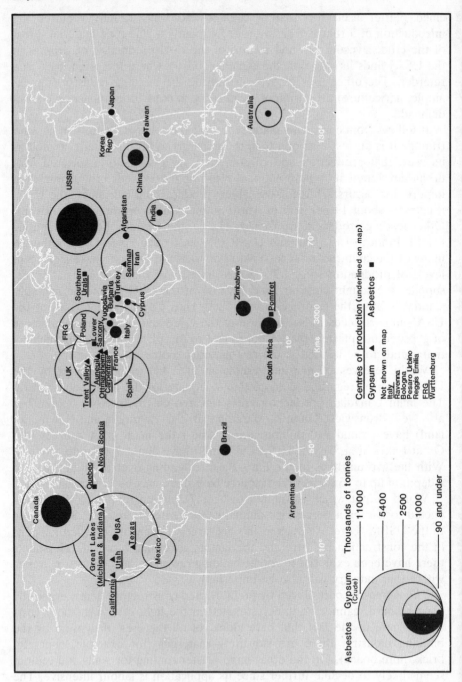

*Fig. 5.3* Gypsum, asbestos: output by country with chief centres of production, 1980. *Source: Mining Annual Review, Minerals Year Book*

*Asbestos*

Asbestos is the name applied to the fibrous forms of a number of minerals which occur in veins of between 0.5 and 1 cm though a few approach 13 cm in width. The most abundant and widely used of these is hydrous magnesium silicate more commonly known as chrysotile; 90 per cent of the asbestos worked is formed from this mineral. Other minerals – amosite, crocidolite, tremolite, anthophyllite and actinalite – may be sufficiently fibrous to be considered as asbestos. They belong to a group of silicate minerals known as amphiboles, but they are brittle and lack the strength and flexibility of chrysotile.

They also are unable to resist attack by acid in the same way as chrysotile but they share with it the virtue of good insulation against high temperatures. Unlike the other construction materials so far discussed, the occurrence of asbestos is much more limited. The primary producers are Canada (24% of world output), the USSR (50%); South Africa (6%); China and Zimbabwe (5% each); Italy (3%) and the USA and Australia (less than 2% each) (Fig. 5.3).

South Africa is the only source of amosite and the prime supplier of crocidolite whilst the USA is the main producer of anthophyllite. There is a substantial international trade in asbestos. Canada, which commands two-thirds of the free market, sells primarily to the USA (that is one-third of its output) and Europe, but with a considerable trade to Far Eastern countries, especially Japan, and Latin America. South Africa, with 12 per cent of world trade, sells most of its output in Europe and Africa. Italy, with 6 per cent of world trade, markets only in Europe. China uses its entire output domestically. The USSR trades asbestos primarily with COMECON, though it is now exporting to France, the German Federal Republic and Japan.

The extraction of asbestos is generally by open pit although African deposits are all worked underground. In the USSR a mixture of both techniques is used. The winning of asbestos creates considerable amounts of waste rock. In the early 1950s on average, 75 per cent of the rock mined passed on to the milling process where the recovery of fibres ran at about 9.9 per cent. The percentage of waste created in the 1970s has risen markedly with only 32.2 per cent of mined rock being milled and a fibre recovery rate of 6.1 per cent. A general downward trend in the richness of the rock bodies worked appears inevitable as the better sources of supply are worked out.

The process of asbestos milling is a complex operation which involves the separation of the fibre from the rock and its classification according to fibre length. Where fibres are very short, special grinding and wet milling processes are deployed to make the material suitable for specific end uses. The longer asbestos fibres are frequently compressed into bags containing about 45 kg of material for onward transmission to manufacturing plant. Because asbestos is a high value commodity as it emerges from primary processing and can easily be compacted there is little association between points of extraction and those of final manufacture. The longest fibre asbestos is primarily used in the making of fireproof cloth for a variety of end products. Much more important, since it consumes about a quarter of total output, is the use of the shorter fibres in the production of cement pipes. Other major uses are the manufacture of flooring products (about one-sixth of total output) and roofing materials and asbestos cement sheet (each in the region of one-tenth). Small quantities are used for thermal insulation in buildings. In total it is probable that about 70 per cent of world production finds its way into the construction industry.

World production of asbestos is estimated at well over 5 million tonnes by the *Mining Annual Review* (1982). The rate of growth in the use of this material appears to be at a relatively higher rate in the developing countries than elsewhere, primarily because of their new construction needs, though in terms of tonnes consumed the USA leads the way.

Demand to the end of the century could show a percentage annual growth rate of 4.1 according to the USBM. However, the construction component within such forecasts must be dependent on factors previously examined in relation to the construction industry and the effects of substitution. In the main asbestos cement products area, there are certainly a variety of alternative ceramic and new plastic materials available. However, prices of plastics could run ahead of those for asbestos since they are products of the petrochemicals industry, while other construction materials, especially wood, could become less competitive in price terms.

In terms of overall future supplies of asbestos some distinctions can be made between the chrysotiles and some of the other rarer sources, and between the various lengths of chrysotile fibres. However, the construction industry is primarily concerned with the former rather than the latter, though some small amounts of amosite are used in fire-resistant marine partition board. Further, it does not use the longer fibred chrysotiles; these are required for spinning and are already in short supply compared with the more readily available medium to short fibred chrysotiles. Thus as far as the construction industry is concerned it appears that demand can be met up to the end of the century from known reserves (standing at 142 million tonnes) even though these may be more expensive to work because of higher waste to valuable material ratios. Thereafter problems may begin to arise if fresh sources of asbestos are not discovered.

However, any discussion of the likely future demand for asbestos cannot be complete without some consideration being given to the health hazards that may result from its production and use, if only because of the publicity this has enjoyed in the 1970s. Problems pertaining to the inhalation of asbestos dust have been recognised in the developed world for many years and in the UK, certainly since 1931 when some regulations were drawn up. However, these were not applied to all occupations at risk nor to all aspects of asbestos manufacturing, ignoring altogether such diseases resulting from it as lung cancer.

In the UK in 1979 the Advisory Commission on Asbestos, set up largely as a response to widespread adverse media comment regarding this group of minerals, recommended banning the use of crocidolite, severely controlling amosite levels in the atmosphere and to a lesser extent, the commoner chrysotiles. In 1983 Professor Donald Acheson, Chief Medical Officer to the UK Department of Health and Social Security, went further in endorsing these recommendations, suggesting a total ban on the import of amosite.

If the developed world ultimately follows these proposed regulations then the production of crocidolite and amosite could be severely affected. However, since these are largely limited to South Africa, a producer which only contributes 6 per cent to the world output of all forms of asbestos, as well as the very small producers of crocidolite, Western Australia and Bolivia, it is not likely that world trade will be especially disrupted. Moreover, much of the South African trade in these forms of asbestos is already with other African countries where health hazards are not a major concern.

With chrysotile, the problems that have been recognised over the past fifty years in the developed countries derive not so much from a need for its total ban, but for higher standards of control from its mining through to its manufacture and usage, and the enforcement of these controls which seems to have been lacking in the UK according to reports presented to the Health and Safety Executive in 1983. In the wake of new regulations concerned with the presence in the atmosphere of chrysotile particles and calls for the rigorous enforcement, some fall in demand from developed countries may be anticipated in the short term, especially where substitution is possible. However, since demand for asbestos is likely to grow faster in the less developed countries up to the end of the century, it may well be that these health problems will not have any great influence on overall output.

The question of asbestos as an environmental hazard is discussed at length by Sandbach (1982) in a companion volume to this series.

### 5.2.2 Phosphate and potash; sulphur; salt; fluorspar; barytes; boron

*Introduction*
As the general introduction to Chapters 5, 6 and 7 made clear, it is possible to group certain minerals according to use inside the broad categorisation adopted.

In this section the minerals have a strong association with the chemicals industry. The development of this form of industrial activity may be looked upon as largely the product of the twentieth century, though it had its origins in the exponential growth of industrialisation in nineteenth-century Western Europe and North America. Only since the First World War has Japan come to the fore and much more recently the rapidly developing countries of South America and the Middle East. The vast range of products emerging from the chemical industries use as their feedstock the minerals referred to in this section which themselves cover a broad spectrum whose ubiquity, general availability and degree of processing are reflected in price. However, it should be added that only the non-metalliferous minerals related to the chemicals industry are discussed here. A limited number of metallics with similar associations appear in section 6.2.4 (Ch. 6).

Minerals with a large range of applications and/or particular significance in the industrial world are dealt with in the following narrative under separate headings, though in the case of potash and phosphate these are considered together because of their common function as fertilisers. However, it should be remembered that many of the minerals alluded to have their own non-chemical uses. While, for example, fluorspar is used to produce hydrofluoric acid and aluminium fluoride, much of it is used directly for metallurgical purposes.

Barytes, though important in the production of barium chemicals, perform a key role in the oil and gas well drilling industry. Soda ash, in some instances the result of the processing of salt by the chemical industry, is itself greatly in demand to meet the needs of glass manufacture.

Conversely, a number of minerals which feature in the chemicals industry are not discussed here. This is not just because they are metallics and not appropriate to this chapter, but because in some instances their utilisation in the chemical industry is very limited or overshadowed to a considerable degree by other uses. For example, of the considerable output of bauxite, around 10 per cent is used in the making of alumina hydrate and aluminium sulphate rather than for aluminium. Although used for metallurgical purposes manganese is utilised to make manganese dioxide and manganese sulphate. Magnesite, important as a refractory, finds some of its output (about 21%) taken in the production of caustic calcined magnesia. Likewise, silica sand, once much used as a refractory and now an essential ingredient of glass, also finds a limited market in the form of soluble silicates in the chemical industry. Both sodium and potassium silicates formed by fusing silica sand with soda ash or potassium carbonate, are used in soaps, washing powders and industrial detergents. Soluble silicate derivates are also used as adhesives, fillers, extenders and absorbants, as well as forming the base for a number of creams, powders and pastes. Then there are the chemicals derived from gas and oil. From these hydrocarbons (discussed in Ch. 7), ethylene, propylene, butane and butadiene are made and used in the production of plastics and synthetic fibres. The refining of petroleum also produces the feedstock from which hydrogen and ammonia are produced.

Finally, there remain a small number of minerals both non-metallic and metallic whose main purpose is to serve the immediate needs of the chemical industry, but whose overall economic importance and levels of output hardly justify the extended treatment given to the minerals discussed immediately below.

*Phosphate and potash*
Two of the minerals in this section share a major role in agriculture and as

fertilisers their significance is further enhanced by the fact that they cannot be substituted for. Along with nitrogen, phosphate and potash are essential to any system of commercial agriculture in that they represent the basic nutrients necessary for plant growth. Indeed, about 80 per cent of all phosphate produced is used as fertiliser and 95 per cent of all potash. Through their application they are largely responsible for the high levels of crop production achieved in developed countries and hold the key to greater productivity in the less developed countries where soil, climate and terrain also happen to be suitable and the appropriate technology is available.

However, their world distribution patterns regarding supplies differ markedly. Of the fifteen producers of phosphate of any substance (Fig. 5.4) the great majority are less developed countries, many of them with a considerable potential for agricultural development. Of the three major producers which currently account for 72 per cent of world output, the USA, the USSR and Morocco, obviously only the last named may be considered as less developed. Given its limited agricultural potential with respect to terrain and climate compared with either of the other two major producers, over 90 per cent of its output is exported primarily to the highly developed agricultural countries of Western Europe, but with almost half going equally to the Eastern European countries and to the less developed nations of the Far East and India. Where the USSR is concerned, it exports phosphate to its Eastern European neighbours but retains 75 per cent of its production for itself. The largest world producer (over twice that of the USSR), the USA, retains only about two-thirds of its output, exporting the rest mainly to Western Europe but also to Japan and Canada.

Only ten countries extract potash and all of these may be considered developed (Fig. 5.4). Two dominate output and between them the USSR and Canada supply 57 per cent of demand; the two German Republics together meet a further 22 per cent with the USA a poor fifth at only 8 per cent. Comparing the free Western market countries with those of COMECON the former is currently meeting 57 per cent of supplies and the latter 43 per cent. Supply patterns as they exist in the early 1980s are however of relatively recent origin, having gradually evolved from a situation prior to the First World War in which Western European producers enjoyed an almost total monopoly of production, to one where they now contribute only about 32 per cent. As would be expected, the demand patterns follow those of phosphate with the USSR and the German Democratic Republic meeting the need of COMECON countries as well as their own. Also the market economy producers meet those of similarly disposed countries but with the developed nations of Europe and North America accounting for three-quarters of total consumption simply because of the advanced state of their agricultural sectors.

Potash is the common term for compounds of the element potassium and is found in bedded underground deposits formed by the evaporation of ancient seas, though it can be similarly obtained through the evaporation of

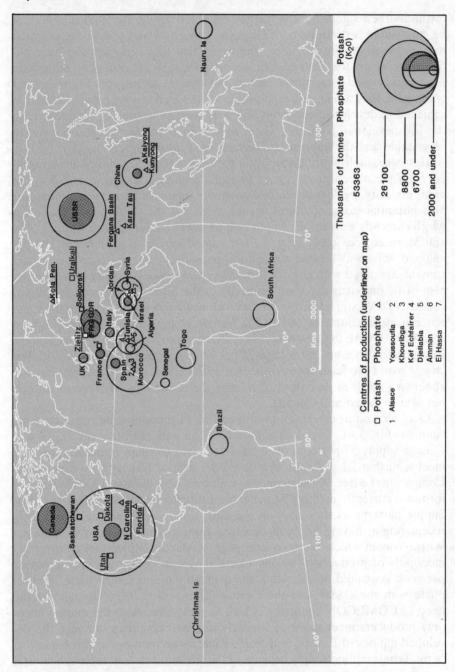

*Fig. 5.4* Phosphate, potash: output by country with chief centres of production, 1980. *Source: Mining Annual Review, Minerals Year Book*

present day salt lakes such as the Great Salt Lake, Utah, the Dead Sea and those in Jordan. Potassium is also found in silicate rocks but the concentrations are generally too low to make extraction worth while. The bedded formations in the Permian basins of New Mexico and eastern Utah and the Devonian beds of Dakota in the USA, which are contiguous with those of Saskatchewan and Alberta in Canada, are the main North American sources. In Saskatchewan these beds are best developed both in area and quality. They contain 25–32 per cent of potassium salts with reserves of nearly 3,000 million tonnes using conventional underground mining techniques. However approaching 70,000 million tonnes of resource may possibly be recoverable by solution which involves dissolving the rock *in situ* underground and pumping the solution to the surface through a pattern of drill holes. Solution mining is already used in the north-east of England on the edge of a Permain basin extending south-east to Germany and Poland and with particularly thick beds in the central section near Hanover, though German Federal Republic mines are contiguous with the main coal-mining area and the German Democratic Republic mines located on the flanks of the Harz Mountains. The USSR deposits occur in three Permian basins; around Berezniki and Solikamsk in an Upper Devonian basin between Minsk and the Pripet River (Soligorsk), and in the Miocene of the Ukraine south of Lvov. In France potassium salts are extracted from the Alsace deposits of the lower Oligocene period; those of Italy from the Miocene deposits of Sicily and in Spain from the thick succession of Tertiary rocks in the Ebro Basin between Barcelona and Pamplona.

Whether the potash is mined in rock form (a relatively easy and inexpensive task because of the comparative softness and non-abrasiveness of the raw material, certainly in its most abundant form, sylvinite), or extracted by solution, or as salts collected from evaporation ponds, the salts are concentrated by dissolving and recrystallising the potash or by flotation and crystallisation. The final product is usually subject to no more than grading according to size and sold in the form of potassium chloride, though there are process variants including potassium sulphate. The relative simplicity of processing results in totally integrated production capacity at all the extraction sites.

Reserves of potassium salts on a world basis are difficult to estimate. Most of the USBM figure of 9,100 million tonnes comes from the bedded deposits discussed above although the figures added to the total to take account of the USSR are based on very speculative information. Further resources might be in advance of 76,000 million tonnes including the currently uneconomic silicate deposits, but leaving out the possible extraction of potash from sea-water where its presence is estimated at 460 ppm.

Phosphate rock (that is a rock containing more than 20 per cent phosphorus pentoxide) can be of sedimentary marine origin and contain a number of phosphate minerals together with a large range of impurities including organic matter, iron and aluminium oxides or phosphates, carbonates of lime or magnesium and fluorine compounds, whilst the rocks themselves are interspersed with clay and sand. These are believed to have been created by divergent up-wellings of sea-water, by deposition in warm currents along eastern coasts and by deposition from water on stable shelves or in continental interiors. It is from such forms of phosphate rock that 80 per cent of the world's production of phosphate comes and will continue to come in the medium term future since they also form the major stock of reserves. Examples of marine phosphate rock appear in North Carolina and Florida. However mineable concentrations of phosphate are also found in igneous rocks as apatite in the Kola Peninsula and in the Kara-Tau deposits of Kazakstan both in the USSR and the Khouribga and Youssoufia areas of Morocco. As yet largely untapped reserves of apatite exist at Palabora in South Africa and Araxa and Jacuperanga in Brazil. Also a small percentage of the world's output of phosphate (about 3%) comes from the Guano derived deposits of the western Pacific Islands.

The working of the sedimentary marine phosphate deposits involves open-pit excavation or strip mining. Once the overburden has been removed by drag line or bucket exacavator, the separation of the valuable mineral and beneficiation is achieved by washing, classification and flotation processes. The igneous deposits require more conventional open-pit hard rock techniques prior to the preliminary processing. Thereafter whichever of the types of phosphate is involved additional processing is needed. For the smaller non-fertiliser market, the phosphate is used to produce elemental phosphorous, electric furnace phosphoric acid and anhydrous derivatives, but for the agricultural sector the phosphate mineral is digested with sulphuric acid to produce phosphoric acid from which the two major phosphate fertilisers, diamonium and triple superphosphate may be obtained.

The fact that prices for phosphate fertilisers and acid are quoted on the basis of shipments from the US Gulf is indicative of two important features of phosphate production. First, that this area has in recent years been dominant in the world export market and second, that exports certainly from the longer established producers are of processed materials, either the fertilisers themselves or in their intermediate state as acid from which the final product may be made. However, as has been noted, export markets are being rapidly built up by other countries whose secondary aim is to move as quickly as possible from phosphate rock production and export to totally integrated facilities which will not only be responsible for mining the raw material but also for processing it.

Thus, Morocco, for example, which has achieved a reasonable export market currently only processes 12–13 per cent of its output but is aiming to raise this to a minimum of 30 per cent in the immediate future. Similarly integrated plant is already in production in South Africa whilst others are being commissioned in Jordan, Iraq, Israel and Syria. In addition, plans to start integrated production facilities where phosphate has not hitherto been extracted are in existence in Egypt, Mexico, Peru and Columbia.

Estimating the overall reserves as well as the resource base for phosphate is not easy. The various types of phosphate rock deposits, as well as their considerably inconsistent levels of phosphorus pentoxide content, create problems where systematic and extensive prospecting has not been undertaken. Moreover rapidly increasing rates of recovery of the basic mineral during processing have made economic not only the reworking of former wastes but also the utilisation of lower grade ores. However, USBM estimates give a total world reserve of phosphate of around 34,500 million tonnes with about half of that figure in Morocco, plus a further 95,000 million tonnes of resource, again largely located in that North African country.

With regard to the future, certainly up to the year 2000, supplies of both potash and phosphate seem assured though in the case of the former the poor spatial distribution of the mineral may become even more apparent as Saskatchewan as a producer becomes ever more significant in world terms. Even with the more widely available phosphate some regional dislocations are probable, especially as the major marine deposits of the USA, particularly in Florida, seem likely to run out before the end of the century.

These forecasts are mainly based on a range of assumptions made by the USBM which support the basic notion of a probable average annual world growth rate in consumption of 3.6 per cent for potash and around 5 per cent for phosphate. However, such calculations are not as heavily based on the likely vagaries of industrial boom or slump and the fortunes of the world economy as are so many other minerals. Here we are dealing with minerals whose output is bound to be strongly determined by the needs of the agricultural sector. Since the world's population will grow and agriculture will need to expand further to meet the situation, a rising demand for phosphate and potash seems assured, unless malnutrition and even starvation are to be contemplated on a massive scale. In spatial terms, however, the growth factors will be more variably apparent. The developed countries have already developed their agricultural potential to a large extent and can sustain high yields. In the case of the USA, for example, UN data indicates arable land to be 20 per cent of its land surface and it is unlikely to expand further to any appreciable extent, whereas the figure in less developed countries is very much smaller at well under half that amount. While the probability exists that the USA has a more favourable climate and terrain, it is all the same reasonable to assume that for the immediate future the conversion rate of land for agricultural purposes will be running at a high rate in less developed countries if they are remotely to approach that figure of 20 per cent. At the same time yields will need to be more than doubled to reach those extant for cereals in the developed countries at the end of the 1970s. All this will be required if the less developed countries are to cope with their levels of population increase (which contrast strongly with the developed countries where population levels are static or falling) notwithstanding any attempts to bring their levels of nutrition into line with the countries of Western Europe and North America. However, all these inputs which indicate probably only modest increases in demand for phosphate and potash in the developed countries but a strongly rising demand in the less developed countries will inevitably depend on the ability of the latter to pay for the import of fertilisers in general. This is especially the case with potash, where, as has been noted, no domestic supplies are available.

*Sulphur*
Because of the importance of the role which sulphur plays in the fertiliser industry – in its sulphuric acid form it is used in the processing of phosphate rock – an argument could have been made for its inclusion in the section on phosphate and potash. However, the fertiliser industry is only responsible for half of total demand. The rest comes, primarily as a requirement for acid (some 90%), from a wide range of industries where it is used as a chemical reagent rather than as a component part of any finished product (Table 5.1). Moreover, other aspects of its production potential, particularly for the future, warrant separate discussion.

Although its working dates back to Roman times, it was not until the birth of the chemical industry in the eighteenth century and its rapid growth in the nineteenth century that the mineral became of major significance. The early forms of its production involved the working of native sulphuric deposits, particularly in volcanic areas where they were derived by the reaction between escaping hydrogen sulphide and sulphur dioxide. Subsequently the rather more ubiquitous sources of pyrite (or ferrous sulphides) became important. These pyritic deposits are mostly of hydrothermal origin. However this left unutilised the elemental sulphur deposits associated with the evaporites anhydrite or gypsum formed by hydrocarbon reduction of anhydrite and probably assisted by bacterial action, a situation which pertained until

*Table 5.1*  Industrial uses of sulphuric acid ($H_2SO_4$)

| | |
|---|---|
| 13% | Plastics and synthetic products (mainly synthetic fibres and textiles). |
| 7% | Paper products (mainly manufacture of wood pulp). |
| 9% | Paints (in production of titanium oxide pigment). |
| 11% | Non-ferrous metal products (mainly for use as a leachate for copper and uranium ores). |
| 7% | Explosives. |
| 5% | Petroleum refining. |
| 2% | Iron and steel production (as a pickling agent to remove rust, dirt, grease from steel products). |
| 46% | Other uses (a wide variety of end uses, including intermediate chemical products, no one of which accounts for as much as 1% of sulphur demand. Also some direct uses of elemental sulphur involved). |

*Source*: USBM (1980)

the end of the last century. It was then that the Frasch method of extraction was first developed for the working of these deposits. This involves the injection, via wells, of large quantities of hot water into the buried deposits containing the sulphur. The sulphur around the wells melts and can be extracted in liquid form of high purity. The Frasch system was quickly applied to the large sulphur deposits associated with the salt domes of Texas and Louisiana in the USA, a country which by 1913 had become the world leader in production and export, a lead which it has not relinquished.

However, whilst production via the Frasch method remains the dominant source of the mineral for the manufacture of sulphuric acid, a basic change in the major sources of sulphur supply is now taking place in certain countries. It is thus to be anticipated, in the case of the USA, to quote the most obvious example, that while this method of primary production currently supplies something under 70 per cent of domestic requirements, the longer term expectation is that 83 per cent will ultimately be obtained from recovered or by-product sources, regardless of cost. Such a situation, strange as it may appear at first glance, is likely to overtake any developed country that has a strong metal mining or hydrocarbon processing industry and a concern for environmental quality. The reason for this is that in the smelting of other metals such as copper, lead and zinc, sulphur is given off as a byproduct. Both natural gas and oil frequently contain sulphur which needs to be removed or recovered while preparing the gas for consumption or in the refining of oil. Although the atmospheric discharge from these operations of such sulphurous gas has in the past been permitted, damage done to the environment by such practices had, since the 1960s, increasingly been recognised and controls have been enforced with growing rigour to reduce their discharge, if not finally to eliminate them in the longer term (see Ch. 8). Technology *does* allow such an ultimate solution, but only at increasing cost. Even so, where the primary producers of metals and hydrocarbons are forced into the position of having to contain and collect their sulphurous byproducts, they will obviously attempt to market these. It is for this reason (plus the fact that anhydrite sources of sulphur will in any case be heading towards depletion within the medium term) that the USBM is forecasting that by the year 2,000 production of sulphur using the Frasch method will have been phased out and replaced by production from the environmentally related sources described above.

The impact on the structure of the sulphur producing industry will be particularly great in the USA with a change from a situation where output is dominated by Frasch production in Texas and Louisiana, to a highly dispersed pattern of much smaller points of production associated with natural gas extraction in Alabama, Florida and Mississippi, petroleum refining in California and the smelting industries associated with copper, lead and zinc mining.

Already the USBM has indicated that recovered sulphur was available from 135 plants located in 28 states. Of these 75 were refineries and 60 natural gas treatment plants. Output was modest at each production point with only three reporting in excess of 100,000 tonnes of sulphur per annum. However their total output, at 23 per cent of total sulphur production, was second only to that of the Frasch industry. By-product sulphur was the third most important source of sulphur, at 6 per cent of total output, and this came from twelve copper smelters and ten lead and zinc roasting and smelting operations, scattered across thirteen states. Even now, pyrite sources of sulphur account for the smallest section of US production, standing at only 1 per cent of the total. Looking at the marketing of sulphur from US sources, the same pattern may be seen with the Frasch industry formerly predominant through the country, having now contracted its markets to its own home territory of the south and eastern states though with immediate access to eastern seaboard ports and thus cheap transport to Western Europe, its export market has held up well. Conversely, the recovered sulphur and by-product sulphur sectors have progressively dominated markets in the western and central states and are pushing into those of the south.

Although the USA contributed around 22 per cent of the world output of over 55 million tonnes of sulphur, it is in its different forms, produced by a large number of countries (Fig. 5.5). None of these predominates as a producer or a supplier to world markets. In descending order of importance after the USA and the USSR, Canada, Poland, Japan, Mexico, France, China, the German Federal Republic and Spain, they account for a further 62 per cent of output. Of the remaining 16 per cent of total supplies, these were produced by a further 57 countries.

World trade in sulphur is considerable with about a quarter of total production thus involved. The most significant export movements were about one-third of Canadian output to the Pacific area, Western Europe and the USA; one-quarter of Polish output to Western Europe and South America; and one-sixth of Mexican output to the USA and Western Europe.

In terms of the future viability of sulphur production there are clearly no problems. Apart from the recovery of sulphur from natural gas and the potential represented particularly by their reserves in the Near and Middle East, Canada and Western Europe, and the opportunity provided for sulphur recovery from petroleum refining especially in the Near East, there are those offered by the association of sulphur with the recovery of copper, lead and zinc (whose reserve and resource base are discussed later). Also there is a potential for further pyrite exploitation especially in Western Europe and the USSR, and there are still large viable deposits of primary sulphur in Mexico, Iraq, Poland and the USSR. Moreover other potential sources of sulphur remain largely untapped; for example the sulphates of anhydrite and gypsum can be processed to form sulphur. That this is largely not done is a matter of cost only. Then there are also the ferrous sulphides associated with deposits of coal also awaiting exploitation. Though small quantities are recovered from these sources in Europe, the potential for co-production output in the rest of the world is very great. The complex organic sulphur

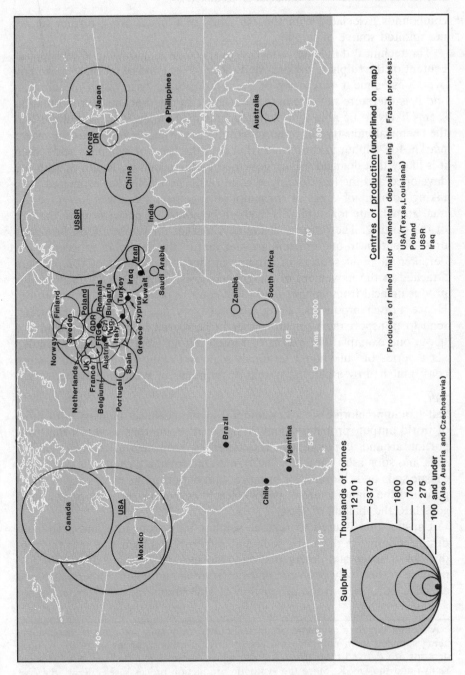

**Centres of production** (underlined on map)

Producers of mined major elemental deposits using the Frasch process:

USA (Texas, Louisiana)
Poland
USSR
Iraq

Sulphur

Thousands of tonnes
—12101
—5370
—1800
—700
—275
—100 and under
(Also Austria and Czechoslavia)

*Fig. 5.5* Sulphur: output by country with chief centres of production, 1980.
*Source: Mining Annual Review, Minerals Year Book*

compounds associated with the tar sands of Canada are a further totally unexploited source of sulphur.

The technical difficulties attached to speculating about the likely sulphur content of the sulphate deposits and those of coal and of oil shale make any overall assessment of reserves or the total resource base rather meaningless and it is therefore not attempted. Suffice it to say that the supply situation seems likely to be able to match demands made upon it by agriculture and the chemical industry. Since no substitutes are available insofar as the major market for sulphur is concerned (that is in the form of sulphuric acid), then it is likely that demand will be governed by the economic health of both the developed and the less developed countries. The USBM has estimated that, taking both major areas of consumption together, a combined average annual growth rate for sulphur is likely to be about 4.6 per cent up to the end of the century. The primacy of the acid market for sulphur does, however, draw attention to one remaining issue: that of the degree of processing undertaken at point sources of sulphur production. Because of the problem attached to the movement of a highly corrosive fluid ($H_2SO_4$), facilities for producing acid from sulphur tend to be close to points of acid consumption. Hence Frasch producers, along with recovery and by-product producers, tend to transport their sulphur either in powdered form or as a liquid. To quote one example, the export of Frasch output from the USA is handled via a series of bulk liquid terminals scattered along the Gulf and eastern coast matched by a parallel system of terminals in Western Europe.

*Salt*

Salt (sodium chloride) is a key mineral in the chemical industry with most of world output, probably about 70 per cent, being thus used. Of that proportion around 95 per cent of it is taken by producers of chlorine, caustic soda and soda ash. These may be termed intermediate chemicals since both caustic soda and chlorine, for instance, find outlets in the processing of wood pulp and the manufacture of paper. Nearly half of the soda ash production* goes into the making of glass.

Caustic soda is used to remove sulphur both from iron and petroleum during refining; both caustic soda and soda ash find their way into the production of soaps and detergents; and chlorine and caustic soda are used to produce respectively polyvinyl and polystyrene and nylon. Other salt based chemicals are used in the production of aluminium and in water treatment

---

* Although largely a synthetic product obtained from salt, soda ash is increasingly being supplied from natural resources (nearly 30 million tonnes in 1980). Natural deposits are derived from trona and are mined commercially in the USA, Mexico, Kenya and the USSR. Since the synthetic production process uses energy, the cost advantage of the natural product has recently become apparent. Though world natural production capacity has reached a quarter of total production, in the USA natural soda ash accounts for 90 per cent of all that used.

and sanitation systems, although there are many other end uses none of which accounts for a substantial percentage of the total.

Salt which has been subjected to basic purification processes is used in the preparation of foodstuffs and as a commodity for cooking in the home. It is also used in the manufacture of cattle and poultry feed. In countries with severe winters one major application is in the de-icing of roads. The largest consumer of salt for this purpose is the USA with demand largely a product of the continentality of its winter weather, its size and its extensive road network. There, the amount used varies from 12 per cent of total consumption to 21 per cent for any given year, depending on the severity of the winter weather. However, salt for this purpose is unrefined and almost certainly mined in block form using conventional underground techniques. This is because this method of obtaining salt is the cheapest and further processing usually only involves a limited amount of crushing and classification by screening. There are, though, other methods of obtaining salt. Where natural brine exists due to the interface of salt deposits and ground water, the saline solution is tapped and/or pumped prior to its direct use in solution form by the chemical industry or the precipitation of the salt by evaporation.

The use of brine wells, developed in the 1880s, involves drilling down to the deposit, the lining of the bore hole, and the pumping of fresh water down it for a prolonged period until a cavity has been produced large enough to sustain the production of saturated brine which is then pumped to the surface. All these extractive techniques are deployed in the working of bedded deposits formed as evaporates in, for example, the Silurian period (those of Ohio, Pennsylvania and New York States), the Cambrian (those of the Persian Gulf and India) or the Permian (those of the USSR, Germany and the UK). The salt domes, such as those of the Texas/Louisiana coastline were formed by geologic pressures acting upon bedded salt deposits forcing them into 'pipe like' intrusions. Another production technique, known since prehistoric times, is based on the solar evaporation of brine. In the modern process however, sea-water is allowed to evaporate in shallow pools to produce a concentrated salt solution. Once the calcium compounds have been precipitated, about 85 per cent of the available salt is allowed to crystallise and compounds of magnesium, bromine and potassium, as well as other elements, are removed. Depending on their end use, the salt crystals are then washed and screened and shipped in wet or dry form. Other natural brines apart from sea-water, may also be treated by adaptations of this process.

The ultimate processing of salt from these natural brines also can be undertaken by artifical heat and is justified only on the grounds of the purity of the end product, most of which is used in the food processing industry, or if no other method of salt production is available. Here the brine is preheated to extract the calcium and magnesium and the liquid pumped into an evaporator. To reduce the amount of heat (and thus energy) required a vacuum is applied to lower the temperature at which the liquid may be boiled away. Salt is ultimately drawn off as a slurry and then filtered, dried and packed.

The favouring of one technique over another or the mix of all methods of salt production largely depends on the circumstances of the individual country of production. In the USA about 55 per cent of production is achieved from brine pumping, 32 per cent is obtained from rock salt and 13 per cent produced by using evaporation techniques either vacuum or solar.

Countries without bedded deposits of salt or other natural forms of this mineral are rare and of the developed industrial powers only Japan has no home-based supply. To avoid the more expensive approach to production required by evaporation techniques in a country that could not use solar methods, it therefore trades in salt, again a most unusual occurrence for a commodity of such relatively low value. However, given its needs as a highly industrialised country, Japan has undertaken a major investment in a lower cost but high quality solar salt complex in Mexico which supplies a large percentage of its needs along with Australia and China. These are the only exporting nations of any consequence, although the Netherlands and German Federal Republic supply salt in modest quantities to their EEC partners. The UK also does the same since the closeness of the salt fields of Cheshire to Merseyside allows the export of the mineral at low cost to many parts of the world. Because of the generally low value of the commodity, its limited transport between countries and indeed its necessary movement within countries, tend to be highly mechanised using bulk containers, conveyor belts, pneumatic tubes screw conveyors or bucket elevators. Its cheapness also means that whilst substitutes for it do exist (for example chlorine can be made from other chlorides apart from sodium chloride) the greater cost of these alternatives means that they offer no real challenge. Only in the case of soda ash has the natural product begun to challenge that made from salt, but then only in that limited range of countries where it is found.

With very large quantities of salt well distributed throughout the world (Fig. 5.6) and with sea-water as the ultimate resource when land-based supplies are depleted, future problems of availability seem unlikely. The current resource with respect to land won salt suggests a further expansion with massive new reserves of 60,000 million tonnes just discovered in the Qinghai Province of China and near Gorky on the Volga in the USSR.

However, the use of non-solar evaporation techniques could ultimately pose a problem for some nations if energy consumption per tonne of salt produced remains for technological reasons at its present high level, or if fuel costs begin to rise steeply again (Table 5.2).

Future demand seems to be tied to factors such as population growth and the gross national product of each individual country – certainly the experience since the early 1950s strongly indicates that this has been so in the past. This is hardly surprising given the strong nature of industrial demands made upon this commodity, its close association with food production and

96

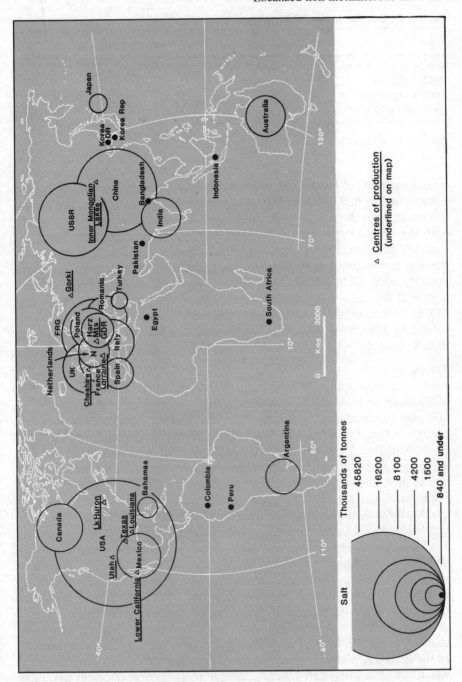

Fig. 5.6  Salt: output by country with chief centres of production, 1980. *Source: Mining Annual Review, Minerals Year Book*

*Table 5.2*  Salt production and energy consumption

| Type of salt | Kw hours per tonne |
| --- | --- |
| Salt in brine | 6.0 |
| Rock Salt | 29.5 |
| Solar evaporated salt | 23.5 |
| Vacuum pan salt | 1230.0 |

*Source*:  *USBM (1980)*

processing and the lack of any viable substitute in many of its uses. Thus the USBM in the mid 1970s confidently predicted a probable average annual growth rate for the world of around 5.8 per cent up to the end of the century with the greatest rises in the levels of demand coming from the newly industrialised countries. However, experience of a world-wide slump in industrial production has resulted in this figure not being realised and more recently reduced to 3.7 per cent. Indeed trends of the early 1980s indicate a growth rate of about 1 per cent per annum. Output seems to be around 157 million tonnes per annum and with earlier predictions of 180 million tonnes by 1985, now unlikely to be realised, the figure could be nearer 170 million tonnes.

*Fluorspar*

This is a common mineral most frequently found in the form of fissure veins along faults and associated with silica, calcite or other carbonates, iron, lead and zinc sulphides and sometimes barytes. The best examples of such deposits are found in north-east Spain, Italy, Sardinia, England, the USA (Illinois) and Brazil. However stratiform deposits in carbonate rocks are also associated with fluorspar, often beneath a layer of clay, shale, or sandstone and again in the same association with other minerals as vein deposits. These occur particularly in the USA, Italy, Spain, Tunisia, South Africa and Mexico. Most fluorspar workings use underground mining techniques, though the bedded deposits are cheaper to exploit than the vein.

Consumption of fluorspar is closely associated with the steel, aluminium and fluorine chemical industries. With a rate of increase in the growth of world crude steel production of something like 65 per cent in the decade up to 1965 and with a further doubling of output in the subsequent ten years, opportunities for an expansion in the market for a mineral which acts as a necessary fluxing and cleansing agent ought to have been good. However, changes in the methods of steel production even further enhanced demand. In the early 1970s steel production relied heavily on the open hearth method which consumed between 1.3 and 2.2 kg of metallurgical fluorspar per tonne of steel compared with the more cost effective basic oxygen method which paradoxically uses a rather greater quantity of fluorspar, 4.5 to 6.7 kg per tonne.

In the ensuing decade the balance swung towards production via the basic oxygen method so that output by this process which was less than 30 per cent of world steel in 1970, had risen to over 70 per cent by 1980. For the electric furnace fluorspar demands were 3.3 to 4.5 kg per tonne of steel produced. Modest savings in the amount of fluorspar used per tonne in the basic oxygen method are however now being made. This has been achieved at the beneficiation stage where, in the production of fluorspar briquettes or pellets, reductions in the amount of mineral used have been achieved through the substitution of aluminium, iron filings, dolomite, olivine, ilmenite, colemanite, borax or pyrohisite, in small quantities as substitute fluxes. Nevertheless little more is likely to be achieved in this direction as the use of such flux substitutes on a larger scale results in a poorer performance compared with fluorspar. It frequently introduces to the process of steel making chemicals which have a deleterious impact on the final product, whilst in terms of quantities of flux required these can be as much as 50 per cent greater.

But the demand for metallurgical fluorspar represents something less than half total demand, the other 55 per cent or so being taken up with the production of the so-called acid grade fluorspar. This consists of 97.5 per cent calcium fluoride compared with the 70–85 per cent calcium fluoride of metallurgical grade fluorspar. Nearly all of this is used to manufacture hydrofluoric acid which is ultimately consumed for the most part either in the process of making aluminium or by the chemicals industry. Here the increased annual demand for aluminium has consistently run at about double that of steel through the 1970s whilst the growth in the use of fluorocarbons (this has increased by 100 per cent over the last decade), particularly for refrigerators and aerosol purposes, has resulted in the consumption of nearly half the output of hydrofluoric acid. Other uses for the acid grade of ore are in the derivation of insecticides, preservatives, antiseptics and in the pickling of stainless steel.

However, where the production of aluminium is concerned the quantities of cryolite (either natural or artificial and manufactured from fluorspar) and the aluminium fluoride required to produce a tonne of primary aluminium have been decreasing, from about 59 to 63 kg of acid grade fluorspar (converted to fluorides) to about 56 kg for each tonne of aluminium smelted in the early 1980s. This has been as a result of considerable pressure applied in the developed countries of the world to reduce the amount of noxious fluorine gases given off to the atmosphere in the smelting process, the achievement of which has meant the efficient recovery and recycling of these gases for use in further processing. But whilst fluorine may be made from hydrofluoric acid, it may also be produced from fluosilicic acid (FSA) which is a by-product recovered from the processing of phosphate to phosphoric acid. In the USA a dozen or so phosphoric acid plants are recovering in total about 50,000 tonnes of fluosilicic acid, a situation which was originally triggered by the need to avoid air and water pollution.

So notable a contribution has the recovery of fluosilicic acid made to the manufacture of fluorine that this has become a main source for all chemical purposes in the USA thus markedly reducing its dependence on foreign imports for this purpose and at the same time maximising the opportunities offered by its indigenous phosphate. Moreover given the importance of phosphate deposits in the USA, the USBM in the mid-1970s went as far as to calculate the reserves of fluorine in the available phosphate rock. Working on the assumption that these deposits contain 3 per cent fluorine, that around two thirds of it is used to make phosphoric acid and that only just over a fifth is recoverable as fluosilicic acid and that the maximum amount of this is in fact recovered, reserves would be around 11 million tonnes. A similarly based calculation for the world as a whole suggests a reserve of about 79 million tonnes. As has been suggested, the USA is not well endowed with fluorspar reserves (a situation shared by its neighbour Canada), although they are relatively widely distributed. Both countries between them import the short-fall of their requirements from the world's largest supplier, Mexico, which currently produces 21 per cent of world output and tends to determine the world price. Other significant producers are the USSR (12%), South Africa (11%), China and Mongolia (9% each), Thailand (4%) and France, UK and Spain (20%) (Fig. 5.7). Although these nine countries produce 86 per cent of world output, production is spread across a further 29 countries, the most important being the less developed countries including Kenya, Tunisia, North Korea, India, South Korea, Pakistan, Taiwan and Burma. The trading links between producers and users are not surprising. The two less developed countries that are leading producers export most of their output: Mexico exports 76 per cent of its fluorspar mostly to North America but also small amounts to Japan and other South East Asian markets and Thailand exports 63% of its output to Japan with smaller quantities going to eastern European countries.

As a developed country South Africa uses about a third of its output domestically but exports about 13 per cent to North America and almost 50 per cent to Japan. The Western European countries mostly supply their own markets. Although small amounts go to Eastern Europe nearly a quarter of their output is exported to North America, a major consumer that is now beginning to receive imports of high quality metallurgical fluorspar from China. The USSR is a major producer but it cannot as a significant industrial power meet its own needs and is a heavy importer of supplies from Thailand and Mongolia.

Most producing countries export both metal and acid grade fluorspar. However, certain countries specialise in exporting specific grades to specific countries. For example, Mexico is the major supplier of acid and metallurgical grade fluorspar to the USA but concentrates on the former, whilst South Africa's contribution to acid grade fluorspar is fourteen times greater than its exports to the USA of the latter. China, on the other hand, only exports high quality metallurgical grade fluorspar to the USA.

Concerning primary processing, beneficiation takes place at the point of

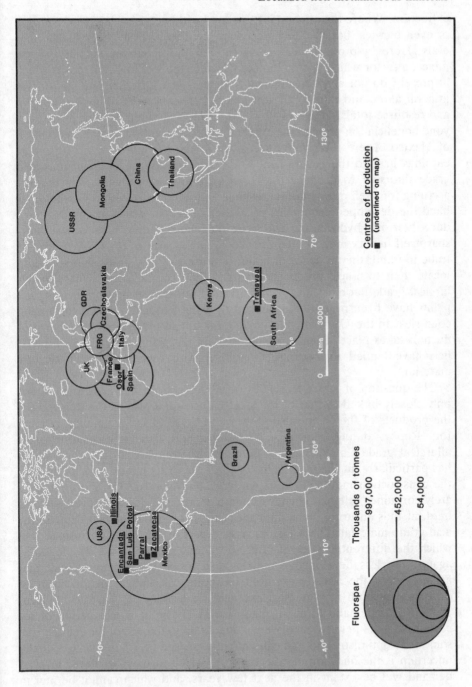

*Fig. 5.7* Fluorspar: output by country with chief centres of production, 1980.
*Source: Mining Annual Review, Minerals Year Book*

extraction. Where the mineral is associated with metals such as lead and zinc or even barytes, flotation techniques are used to sort out the valuable minerals. Here by-product sales contribute to the economic viability of the mines, a factor which could ultimately lead to the working of deposits which at present do not contain sufficient fluorspar to justify their mining for this mineral alone and thus to an adjustment between overall fluorspar reserve and resource totals. However, few less developed countries have moved beyond beneficiation prior to the export of fluorspar with the notable exception of Mexico. There the significance given to the product in terms of export earnings has led the government to develop the use of its indigenous acid grade fluorspar to produce the more expensive hydrofluoric acid thereby accruing for itself more internal revenue and job opportunities. On the other hand the developed economies of Western Europe, which in any case produce their own hydrofluoric acid from indigenous fluorspar, do export fluorspar itself in very modest quantities to countries such as the USA. With this trade in mind, the tendency of major importers of acid grade fluorspar is to locate their own acid producing plants at their coastal ports. Where metallurgical grade fluorspar has been fabricated into pellets and briquettes such plants have been primarily associated with the major developed consuming countries. In the USA where a limited amount of indigenous fluorspar production takes place, companies undertaking the manufacture of such products have tended to associate themselves with sources of the basic raw material.

The question of future levels of overall demand for fluorspar remains not only closely linked to that of the steel industry but also to the demand for the products of the chemicals industry and those of the aluminium fabricators. Reports dating from the early 1980s suggest that the demand for metallurgical grade fluorspar has suffered more from overall recession which has particularly hit the steel industry (because of its major association with the construction industry and expensive capital goods). The more immediately consumer oriented chemicals and aluminium industries have been less hard hit thus ensuring the continuance of a strong and buoyant demand for acid grade material. These factors remain part of the wider environment in which the differential rates of future economic growth in the developed as against the less developed countries have to be taken into account and against which the already explored technological changes in the levels of utilisation of both metallurgical and acid grade fluorspar have to be seen. The USBMs demand forecast, made before the most recent period of recession, of a 4.4 per cent increase per annum up to the end of the century was somewhat optimistic and was subsequently revised to 3.5 per cent. Equally uncertain is the question of reserves from which to meet whatever level of demand will be extant in the next few years, and which cannot be met in other ways (that is the use of fluosilicic acid).

Although the indigenous supplies of the current major consumer seem at 0.86 million tonnes destined to last only a little over 20 years at 1980 rates of extraction (unless submarginal resources can be converted to reserves), rapid strides have been made in identifying new resources on a world basis. In the first half of the last decade the resource base was increased by some 50 per cent. In the period since then each year has seen further major additions not only to the resource base but also to reserve totals. Spectacular increases have occurred in China and Mongolia (countries whose ideologies the USA might view with circumspection if it had to rely on imports from them), as well as in Mexico and South Africa. Indeed total resource estimates now stand at around 376 million tonnes with 80 million tonnes of reserves. No immediate problems are therefore foreseen regarding fluorspar availability, nor are there likely to be major disruptions of world trading patterns.

*Barytes*
A heavy relatively inexpensive industrial mineral comprising primarily the chemical barium sulphate, barytes is sometimes found in association with the ores of lead and zinc as well as fluorspar. Like fluorspar, it is to be found as a vein deposit or as a bedded sedimentary deposit or, most frequently, as a residue replacement deposit. It may be worked by standard opencast or underground mining techniques with the ore then undergoing crushing, grinding and beneficiation by flotation. Only if the barytes is of very high grade, containing up to 96 per cent barium sulphate is it merely crushed prior to shipment, but in order to be worked at any deposit it is required to have a minimum 92 per cent barium sulphate content and specific gravity of not less than 4.2. Although barytes has a number of important uses the current demand for the mineral is governed almost entirely by its employment in the oil and gas well drilling industry and the level of activity therein. Around 90 per cent of world production is currently used in this context where, in a finely ground state, it is required to control the density of the circulating drilling fluids to provide lubrication and to prevent blow-outs. Its quality of inertness plus its stability and high specific gravity make barytes ideal in such a role.

Considerable amounts of high quality barytes are also used for making barium chemicals including barium chloride (used in ceramics production), barium carbonate (for which the glass industry has a demand) and barium hydroxide (where it is used by the electrical industry). A precipitated form of the barytes known as Blanc Fixe finds its way into the production of paints and colours and into the preparation of a compound used for the medical examination of the human digestive tract. In the latter case, its resistance to X-rays enable soft tissue to be revealed during radiological examination. This same quality makes it a suitable mineral for use in concrete or plaster in any building where X or Gamma radiation needs to be contained. Barium sulphate is also used to derive strontium carbonate and strontium sulphate. In the carbonate form it is employed in the production of glass for television display tubes, whilst as a nitrate it is used to add brightness of colour to fireworks and distress flares. Strontium chemicals also appear in the manufacture of ferrites for ceramic permanent magnets and in greases, soaps and pharmaceuticals.

It was in response to the demands of the oil and gas industries that the consumption of barytes rose sharply over the 1970s to a total of 7.26 million tonnes in 1980 with the leading producers being the USA (26% of total output), Peru (6%), India, Thailand and Ireland (each over 5%), Mexico (over 4%) and Morocco, France and Turkey (each over 3%). Italy and Chile also produce significant amounts. The USSR is also known to be a major producer (over 7%) and recently China has greatly increased its production of high quality ore (over 7%) to meet its own drilling needs and for export (Fig. 5.8). Recent levels of international trade have though been largely influenced by the desire of the USA to find new sources of hydrocarbon energy and by its embarking on a massive exploration programme.

Not only were indigenous sources of barytes, particularly those of Nevada, Missouri and Arkansas, under pressure to meet immediate needs but 1.8 million tonnes of imports were brought in from new sources of supply (China, Chile and India) as well as from the more traditional sources such as Peru, Ireland and Morocco. The other major area of demand has been the North Sea. According to the *Mining Annual Review* (1982), it is expected that with exploration continuing at a high level not only will Irish sources of supply be favoured but an added fillip will be given to UK production. These two examples of regional growth in demand are representative of the overall market for barytes where demand can fluctuate spatially depending upon the changing location of oil and gas exploration.

Since interest in the search for hydrocarbons must grow, it is possible to foresee the future of barytes in similarly expanding terms, perhaps with a demand growth rate of 4 per cent per annum through to the 1990s and beyond. Certainly the only possible substitutes in the major demand area are unlikely to be taken up. Iron ores could fulfil similar functions but are abrasive and unpleasant to handle. Celestite has a lower specific gravity and costs a good deal more.

At the end of the century the exploitation of oil shales and tar sands as major hydrocarbon sources could begin to have an adverse effect on the demand for barytes, one which, as things look in the early 1980s, could hardly be expected to be mitigated by the demands of the chemicals industry which remain and are likely to remain small by comparison. Ample reserves of barytes are to be found available within the territory of the major consumer, in the other producer countries mentioned above as well as in several more, giving a world total of around 127 million tonnes of reserves in a resource base probably eight times larger.

With no specific power bloc or economic trading group in a position to control supplies, barytes as a mineral does not seem likely to present a problem in terms of availability during the remainder of this century.

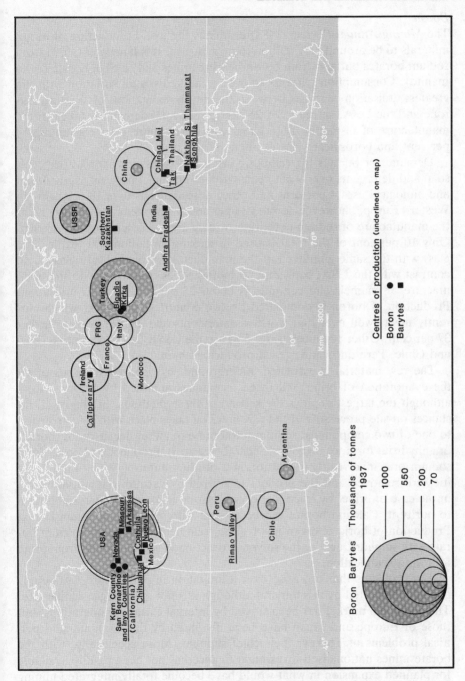

*Fig. 5.8* Boron, barytes: output by country with chief centres of production, 1980. *Source: Mining Annual Review, Minerals Year Book*

*Boron*

The *Mining Annual Review* (1982) estimates the world production of boron minerals to be around 2.71 million tonnes in 1980. It is largely obtained from sodium borates but also from boron-bearing brines and calcium borate (colemanite). Consumption patterns showed a marked spatial variation with the greatest utilisation contrasts between the developed nations of Western Europe and the USA. In the USA 25 per cent of total demand is taken in the manufacture of insulation fibre glass, with textile fibre glass consuming 14 per cent and borosilicate glasses about 12 per cent.

Demand for borates for cellulosic insulation takes just over 12 per cent, soap and detergents over 9 per cent, vitreous enamel 4 per cent, agricultural and biological uses 4 per cent and other miscellaneous uses 18 per cent. In Western Europe, however, around 33 per cent of borates consumption is for the manufacture of sodium perborate, widely used in detergents as a bleach. Only 10 per cent of the total is used in making insulation and textile fibre glass with the same employed in the manufacture of borosilicate glasses. In contrast with the USA, consumption by industries concerned with the manufacture of enamels and ceramics is about 25 per cent of total utilisation. Production of boron is dominated by two countries, the USA, which currently meets well over half world requirements and Turkey with a further 37 per cent. Other producers (Fig. 5.8) are the USSR (8%), Argentina (2%) and Chile, Peru and China (each producing about 1%).

The raw material is extracted underground, by open pit methods or by the evaporation of brine. All three techniques are used by US producers although the largest deploys the second of these methods. In almost all instances on-site processing of the raw material takes place with its conversion to one of two compounds; sodium tetraborate in either decahydrate (borax) or anhydrous form, or orthoboric acid (boric acid). Production of borax from sodium ores involves their solution in a hot liquid during which the impurities are removed and the product recovered by vacuum crystallisation or, in the case of brine, by fractional crystallisation followed by recrystallisation to purify it. Calcium borate can be treated with soda ash to obtain borax. Production of boric acid can involve the treatment of borax or calcium borate with sulphuric acid or the solvent extraction of soluble borate from brine following acidification.

The problems that can arise over a limitation in point sources of supply are well illustrated by this mineral although it is not of strategic significance. The major US market has been served by internal extractive operations, but those of Europe and Japan have been disrupted by the economic and political problems of Turkey, their chief supplier. More specifically, with the borate mines nationalised investment finance has not been readily available for planned expansion in what would have become totally integrated mining and processing complexes; thus progress towards this end has been very slow. A greater degree of longer term stabilisation may come into the market if other producers are able to make a more substantial contribution to world supplies.

China looks most likely with as yet unquantified reserves in the lakes of the Quaidom Basin. Should shortages be avoided, the value of the mineral as an insulation product should ensure a market growth to the end of the century of around 3.5 per cent per annum on a world basis.

### 5.2.3 China clay (kaolin) and ball clay

Both china clay and ball clay are related minerals which make up a group whose value and importance are distinguished not by their chemical qualities so much as their physical characteristics. The significance of china clay lies in its whiteness (due to its high percentage of kaolinite), its rheology and its particulate size distribution, with the nature of the processed product tailored to emphasise one or more of these characteristics in relation to a specific end use. These include the manufacture of paper (which consumes about three-quarters of total output in a developed country such as the UK); the production of ceramics (which consumes 17 per cent of output) and where products such as bone china can contain as much as 50 per cent china clay; in a much more limited market as a filler or an extender in the making of paint, rubber, plastics, cosmetics, lubricants, fertilisers, insecticides, dyes, polishes and so on; and as a specialised refractory. On the other hand, ball clay, though containing a significant proportion of kaolinite, also contains a certain amount of organic matter. This, plus the fineness of the particle size, gives to the ball clay a plasticity essential to its use by the ceramics industry where about 60 per cent of output is used to make earthenware, tableware, wall and floor tiles and sanitary ware. Although ball clay is also used as a dusting agent to prevent the caking of fertilisers during storage and application and as a filler in the rubber industry, its other significant use is in the bonding of pre-fired materials to produce blast furnace bricks, in kiln furniture and in refractory mouldables.

Both forms of clay, together with all the others such as the commoner clays used by the construction industry as well as the refractory clays, are the product of the mechanical and chemical breakdown of parent rocks. All contain a plastic component (mostly the clay content) and a non-plastic portion (consisting of quartz, micas, feldspars and/or iron oxides). All can be designated as either residual or sedimentary clays. The former are those that occur in the same location as they were originally formed. The china clays of the UK were created *in situ* as the result of the breaking down of the feldspar in the local granite under the influence of active vapours rising from the cooling rocks. This led to the production of a hydrated silicate of alumina, kaolinite, intermixed with the two other main constituents of granite, mica and quartz. The ball clays in the UK are sedimentary in origin and since they have been transported by water action, the coarse fragment of the kaolinite has been separated out and deposited earlier and separately from the finer deposits which now form the local clays of the Bovey Basin of Devon and the Poole/Wareham area of Dorset. The process of transportation has however, enabled other impurities to be added whilst the original fine micaceous fraction remains.

As with most clays, china clay and ball clay are mined from open pits using modern surface mining equipment such as power shovels, draglines and front end loaders. However, in the UK, china clay is extracted from the pit by hydraulic mining (using high pressure hoses) and pumping. Classifiers are then used to separate out the quartz fraction and hydro cyclones the micaceous residues from the china clay. These unwanted minerals exist in the parent rock body at an average ratio of seven parts to one of kaolin (though in the UK the ratio may be as high as 8 or 9 to one). In other areas such as the USA, where hydraulic methods of extraction are not used, the clays may be dried, then pulverised before the coarse fragment is removed with the aid of screens. The ultimate processing of china clay will be dependent on the end product but additional separation and grinding techniques may well be employed. Brightness may be improved by bleaching, ultra-flotation and calcination. However, some of the most recent developments have been aimed at improving the whiteness of the end product through the use of magnetic separators to remove all traces of iron and titanium. Moreover with a view to reducing the amount of energy used in making the end product, improved dewatering and drying techniques using mechanical means have been introduced.

Ultimately the china clay may be supplied to the consumers as a powder, in lump form, as dustless beads, as moist filter cake, as extruded pellets or in the form of slurry. The last mentioned development has certainly been the major innovation with transport in this form (up to 70% solids) achieved by road tanker, train or vessel. However the most important market remains in the area of the dried product.

Just as the blending of china clay from several different pits is a necessary prerequisite of uniformity of product in relation to a specific market, so it is with ball clay which tends to occur in separate seams each with a varying mica and kaolin content. However, unlike china clay, if the open pits go below 30 metres or so, the material then tends to be worked underground. Moreover, because the value of ball clay is about half that of china clay and the costs of transport thereby even more crucial, the end product tends increasingly to be de-watered (to save bulk and weight) and carried to its market in a pulverised or dried pellet form rather than in its more traditional state, that of lump or shredded clay.

As with most of the commoner clays, there is a broad spatial distribution of china clay. About 34 countries are known to have an output of the refined product, nearly half of which are developed countries (Fig. 5.9). However, whilst about a quarter of the world's refined output comes from major producers such as the USSR, Czechoslovakia, the German Federal Republic,

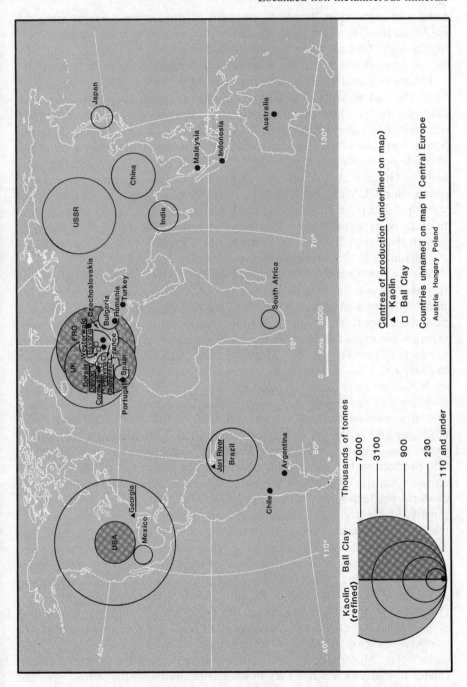

*Fig. 5.9* Kaolin and ball clay: output by country with chief centres of production, 1980. *Source: Mining Annual Review, Minerals Year Book*

India and since 1976 Brazil, 60 per cent of production emerges from the USA and the UK where deposits are not only unusually large but are of very high quality. Outside these last mentioned countries, for instance, only Brazil produces china clay of the specification demanded for paper coating.

Production and trading patterns can therefore be simply categorised. First, the vast majority of small scale producers supply their own indigenous markets (except where premium grades of china clay are concerned) and will increasingly do so, particularly in the case of the countries of South America, South East Asia and Australia. Second, increased production surpluses above home needs occurring in the USSR in the early 1980s, Czechoslovakia and the German Democratic Republic seem likely to continue to be traded only within COMECON. Third, the large surplus output of the two major world producers and consumers, the USA and the UK, seems likely to continue to be exported to the other major industrialised countries of the West though the paper filler market is wider still. Certainly both major producers are fortunate in the location of their china clay deposits.

The St Austell and Lee Moor workings of the south-west of England and the Georgia and Carolina deposits of the USA are not only very close to sea ports, but are well placed in terms of sources of demand. Thus while UK producers export over 80 per cent of their output to adjacent EEC countries, Georgia can export to Western European countries with transport costs little in excess of those of the UK. Moreover, the extraction of china clay in both the UK and USA is dominated by a small group of firms whose large scale extractive activities, product specialisation (achieved through sophisticated blending and processing, and ownership of the means of bulk export using specialised vessels) all lead in the direction of low production costs, high efficiency and giving the consumer exactly what is required at a competitive price. The largest of these firms, English China Clays, not only dominates UK production but has considerable interests in the USA and other producing countries of Western Europe. It alone accounts for 3 million tonnes of world production per annum out of 18 million tonnes of refined product. Finally, of the developed countries, Japan appears to be increasing its links with the new major producer, Brazil, which also exports to Europe, the productive capacity of which has been otherwise developed by US interests.

The production of ball clay is somewhat restricted in world terms (Fig. 5.9). Although estimates have been prepared which indicate the leading producers as the German Federal Republic (3 million tonnes per annum), France (1.5 million), the UK (1.4 million) and the USA (0.8 million), the *Mining Annual Review* considers that the figures for France and Germany contain large tonnages of clay quite unrelated to the manufacture of ceramics, the main use of ball clay. Moreover, the ball clay component of all the countries named relates to material which is very inferior compared with the output from the two UK deposits of Wareham/Poole and the Bovey Basin. Czechoslovakia has ball clay resources of equivalent excellence but there are no good data regarding the amounts extracted each year. Only in the case of the ball clay deposits of Kentucky, Tennessee and Texas in the USA is there a complete picture of the amounts produced and confirmation that these are from deposits of equivalent quality to those in the UK. However, in terms of export markets, the long overland freight routes, compared with their UK counterparts, make the production of these clays, except for home consumption, uncompetitive. Imports from the UK have penetrated most of the states along the US eastern seaboard.

The UK is far and away the leading producer and exporter of high quality ball clay for use in the manufacture of whiteware products. The largest single deposit is that of the Bovey Basin providing two-thirds of UK output. As with china clay, production is dominated by a single producer, Watts Blake and Bearne, which control 55 per cent of UK production. Over 70 per cent of all that is produced in Devon and Dorset is exported to around 40 countries in South America, Africa, and Asia together with Canada and the USA. However, Western European countries are by far the most important market taking around 90 per cent of all exports.

The limitations in the number of supply sources of ball clay could present problems for the future. If anything like the world increase in demand forecasts estimated in the mid 1970s by the USBM as between 2.8 and 3.8 per cent per annum were to be established in the next few years, production levels would soon become difficult to sustain unless extraction techniques could be improved.

As for china clay some demand forecasts have indicated a growth rate for individual countries around that of their increase in GNP. However, it should be realised that in the most highly developed countries an increase in the use of china clay is likely to be somewhat more constrained than in others less favoured. In the USA it could be argued that the use of this mineral in the major market as a paper filler and as a coating for high quality papers and cartons, has reached saturation levels and only an increasing population is likely to have much impact by way of further stimulating an increase in demand. This would contrast with West European countries where the high gloss coating market for both board and paper has yet to be fully exploited and with the less developed countries where it remains totally unexploited. Concerning the less developed countries, increased levels of literacy offer unparalleled opportunities for the stimulation of demand for books and journals and thus for paper fillers and coatings. However, one factor which is likely to have world-wide significance in stimulating demand is that the fillers which have made up something like 30 per cent of the value of paper for around 150 years are between six and ten times cheaper now than paper fibre. There is therefore an across-the-board incentive for a further increase in the use of fillers at the expense of fibre. Some substitution possibilities have been attempted with Imperial Chemical Industries developing a synthetic alternative filler, but these have so far not proved competitive.

Whatever demand forecasts may be posited for the major china clay mar-

111

ket, it is certain that whilst the sources of supply are likely to remain limited in number, particularly for the higher grades of product used by the paper industry, china clay resources are unlikely to be worked out in the foreseeable future. Those of the USA and UK together amount to over 1,000 million tonnes according to the USBM. In so far as can be ascertained from the major producer, the estimated total of over 200 million tonnes of resource available in the UK remains a very conservative approximation since none of the producing clay pits has even been worked out. Although there are plans which predict the completion of the current working at Lee Moor around the year 2030 (assuming an output of 0.634 million tonnes per annum), the latest development plan for St Austell foresees the currently remaining 80 per cent of production requirements met by the workings in this area over roughly the same time span. However, the aim of this plan is essentially to address the problems of land requirements for the disposal of quartz and micaceous wastes (see Chs 1 and 9); there is no question of the resource being totally depleted in some fifty years' time.

**CHAPTER 6**

# Key industrial minerals – metalliferous minerals

## 6.1 Ferrous

### 6.1.1 Iron

The idea of iron ore's pre-eminence amongst all the minerals extracted from the earth may be sustained on two counts. First, it is the fourth most abundant rock-forming element, comprising about 5 per cent of the earth's crust. Even though only a small part of this has subsequently been concentrated by sedimentary, igneous or metamorphic processes into deposits with an average iron content of up to 68 per cent, it is spatially ubiquitous (Fig. 6.1) and available in such quantities at its many point sources as to present no foreseeable problems in relation to future demand. Second, since the last century iron has been the basic ingredient of steel, the most dominant metal in the process of industrialisation. This is simply because of its strength, its capacity to be successfully alloyed with other metals, its recycling potential and its low cost which is itself in no small way due to the widespread availability and abundance of iron ore, as well as the relative ease with which it may be mined and processed.

Its inexpensiveness, around $200 per tonne in 1980 (pig iron), can be noted by comparison with two other widely used metals; aluminium is over seven times and lead well over three times more costly per tonne of refined product. Not surprisingly, iron ore accounts for 95 per cent by volume of all metals currently mined.

The ore itself consists of iron oxides, the mineral forms of which are magnetite, haematite, geothite or limonite. It may also be obtained as siderite (from iron carbonate) or pyrite (from iron sulphides). Concentrations of iron ore are most frequently found in sedimentary materials of pre-Cambrian age. These are mined extensively in the USA, Canada, South America, Africa, India, Australia, the Far East and the USSR. Such deposits, usually consisting of magnetite or haematite and quartz plus some iron sil-

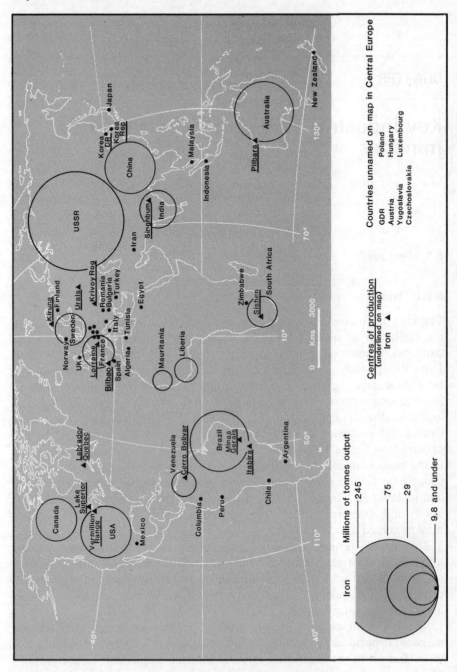

*Fig. 6.1*  Iron ore: output by country with chief centres of production, 1980.
*Source: Mining Annual Review, Minerals Year Book*

icates and iron carbonates, can be between 30 and 600 metres thick and are often exposed over considerable areas but they frequently suffer from complex folding. These formations normally contain 20–40 per cent iron but they may also be locally enriched to high grade deposits of haematite by oxidation and the removal of silica. The oxidised or residual ores with 50–60 per cent iron ore are worked in the Lake Superior area of North America and the Krivoy Rog district of the USSR whilst the residual or replacement ores, 64–8 per cent iron, are mined in the Vermillion range of Minnesota (USA), Minas Gerais (Brazil) and Sishen (South Africa).

Other sources of iron ore are the sedimentary deposits of the oolites of the Palaeozoic to the Cretacious age. These ironstone formations are areally extensive but they are usually less than 15 metres thick and contain between 20 and 40 per cent iron. Although their high phosphorous content causes processing problems, such deposits also often contain calcium carbonate ($CaCO_2$) and are self-fluxing. Thus they are still mined extensively in France (Lorraine) and to a lesser extent in the German Federal Republic and the UK. In pre-Cambrian rocks, concentrated deposits of magnetite of igneous origin are also found which may contain as much as 65 per cent iron ore. These were formed either by the segregation of magnetite crystals in magma and by the injection of a magnetite rich fluid into the surrounding rocks, or by the replacement of rock such as limestone by magnetite as a result of contact with intrusive igneous rock. The latter type are usually associated with the Jurassic or younger intrusives of the western USA and Mexico. The former type, tend often to be associated with titanium (for example those of Lake Sandford, New York State), but they can also be high in phosphorous such as the high grade magnetite ores mined at Kiruna in Sweden. These Swedish ores come from the largest deep mine in the world but are now a rare example of this form of extraction as far as iron ore is concerned. With the working out of the highest grade deposits which merit such treatment, the overall high operating costs and limited production capacity of underground mines has steadily reduced their ability to compete in a world where there are ample bodies of ore close to the surface. Although these are lower grade, the low cost of the open-pit methods has come to dominate production with nearly 90 per cent of output realised in this way in 1982.

The opening up of the lower grade deposits has, however, occurred at the same time as a rising demand for ores of higher iron content and more uniform chemical composition and physical structure. For this reason processing has had to become more sophisticated and in particular the means available for beneficiation before shipment.

Whilst high grade ores may be crushed and screened, lower grades need further extensive treatment by gravity, magnetic or flotation beneficiation processes to produce concentrates before sintering or pelletisation prior to shipment from the point of extraction. The sinter technique involves mixing ore fines with coke and heating, thus causing the ore particles to fuse together. An average iron content for sinter might be about 54 per cent. Pellets are made by mixing small amounts of bentonite with finely ground concentrates and about 10 per cent water. The pellets of about 1–1.5 cm in diameter are then hardened in a gas- or oil-fired furnace, with each containing about 62–5 per cent iron. The great advantage of pellets is the increased efficiency of smelting that results from their use in blast furnaces with uniformity of shape, strength, porosity and composition all promoting a better rate of reduction. Moreover, they are less liable than sinter to deterioration during shipment.

In spite of all these advantages world production of pellets was estimated in 1980 to be only about 190 million tonnes. This amount was 30 per cent below the capacity of existing plant and followed a period of fourteen years up to 1974 when the annual increase in the production of pellets had averaged 18 per cent. The reason for this fall was the cost of production particularly in oil fired plants after the massive price increases of 1974 and 1979 with thermal capacity requirements up to 800,000 Btu per tonne for magnetite concentrates and about 1 million Btu per tonne for haematite concentrates. Those located in developing countries, where their role in locally adding value to mineral output can be of great significance, were particularly badly hit. Of the twenty plants engaged in pelletisation, by 1980 over 80 per cent of production came from the developed nations of the USA, Canada, the USSR, Australia and Japan in that order.

As for the location of pelletising plant, the tendency has not necessarily been towards their construction close to the iron ore mines. Port locations have been chosen whether pellets are to be made from imported ore or exported as well as at the smelter. Any ideas concerning the optimum location of such plant can only be established through the study of specific cases. Although as a result of handling low grade ores it may be necessary to site the pelletiser close to the mine to minimise transport costs and facilitate tailings disposal, these problems have been overcome via the use of low cost slurry pipelines connecting mine sites with ports where, once the pellets are produced, they may be loaded directly into the holds of bulk carriers. Pelletiser locations at the smelter allow the purchase and blending of ores from several sources although usually at the expense of increased transport costs. However it remains true that in spite of the variations referred to above, most pelletisers are constructed close to iron ore mines, advantage thus being taken of the low outlay and general convenience to be found in the bulk transport of pellets by rail or ship. Canada and Australia provide good examples of plant so located. In contrast, Japan, the Netherlands and the UK have smaller plants handling mainly imported ore which are pelletised prior to their use in blast furnaces. They take maximum advantage of differences in the prices paid for ores from various sources and the security of supply offered at times of ore shortage.

The primary destination for all forms of iron ore is that of the steel producing plant. The manufacture of this product (shown by country for 1980 in Fig. 6.2), traditionally has involved blast furnaces and oxygen blown convertors. Due to the large scale of production required for efficient working and the high initial capital outlay, such plant tends to be the province of the developed countries. A typical expenditure in the construction of a plant of 7,000 to 10,000 tonnes per day is around $300 million. However, a new method of steel production known as the direct reduction process has eliminated the need for blast furnaces as well as high levels of output. Small direct reduction plants with an annual output of around 100,000 tonnes can be purchased at a sum well within the means of the developing countries thus enabling them to enter the steel production business and, by additional processing, add further to the value of their own indigenous mineral resources. By 1980 direct reduction plants were completed or under construction in at least twelve countries. World production (outside the USSR, Eastern Europe and China) from this process was around 9 million tonnes (1.35% of world steel output) with Latin America responsible for 45 percent of this total. The *Mining Annual Review* believes that by 1985 total capacity will rise to 24 million tonnes (or 2.5% of the forecast of world steel requirements) from 41 direct reduction plants 18 of which are expected to be in Latin America.

As far as the less developed countries are concerned steel production in both Brazil and India is expected to rival the output of many of the developed countries, and China plans to increase its output by over 60 per cent between 1980 and 1985. This situation will inevitably further change the world map of the interrelationships between suppliers and consumers, one already considerably modified by the experience of the 1970s. During that period steel-making plant of the conventional type in North America and Western Europe found itself increasingly unable to compete with countries such as Japan. In the USA, Canada and the UK the ability of the steel trades unions to secure high wage settlements led increasingly to the uncompetitiveness of their steel producers and this in turn discouraged investment in new capital equipment. However, by the early 1980s even Japan found itself competing with even lower cost new steel producers such as South Korea where wages of $100 a month prevail in steel-making plant. The short term response both by the US and the EEC to their declining steel industries is that of import control. The EEC plans to set steel output targets in response to forecast levels of import demand and to impose compulsory minimum prices for all imports of steel to the EEC.

These structural changes in steel making capacity have had an impact on iron ore producers and their traditional markets. With the decline of ore imports by the USA and the European Coal and Steel Community (particularly the UK), a decline in exports by the related ore-producing countries (Canada, Sweden, Liberia and Venezuela) of between 15 and 20 per cent has been experienced in a period as short as one year (1979–80) and not totally offset by the development of new markets though these continue to grow apace in Asia, Africa, the Middle East and Latin America.

117

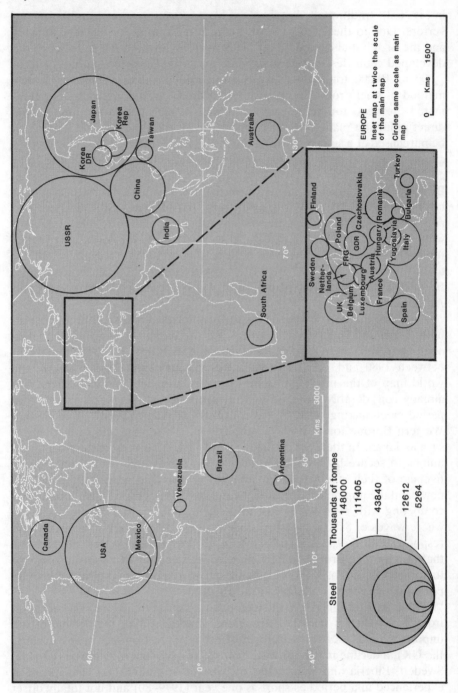

Fig. 6.2   Steel: output by country with chief centres of production, 1980. *Source: Mining Annual Review, Minerals Year Book*

Only ore producers serving the export markets of Japan and South Korea where steel production has continued at a high level and in the case of the latter to expand, have not been so affected. Both Australia and Brazil had thus further increased their exports of ore to record levels in 1980. However, Brazil was also supplying its own steel industry with ore as was Mexico, Argentina and Venezuela. Steel production in all three of these last named countries expanded in 1980 as a result of investment in plant using the direct reduction method.

Because iron ore is a relatively inexpensive mineral, the cost of transport is a significant part of the cost of the ore delivered to the consumer. The relationship between suppliers and consumers tends therefore to be that of the closest proximity with shipping the main transport mode. For this reason US East Coast steel manufacturers find it cheaper to import seaborne foreign ores than use Lake Superior supplies that must be transported by lake carriers plus rail. The transport mark-up of Superior ores so moved can be as much as 50 per cent of their estimated delivered cost.

Superimposed on the changing relationship outlined above are the patterns of supply and demand for iron ore, and ultimately the steel manufactured, by the normal range of economic factors. In the early 1980s the picture was inevitably one of slack demand though it represented probably only a downward fluctuation within an overall upward growth trend albeit now no longer exponential. Even so future predictions about demand up to the end of the century are essentially speculative. The most reliable estimates concerning iron ore requirements to the year 2000 come from the USBM. Assuming world demand will grow at about 3 per cent compounded annually and that the ratio between raw steel production and iron ore input that currently exists will be similar at the end of the period, it calculates that the total figure will be in the region of 1,000 to 1,300 million tonnes by the year 2000. These estimates cannot, however, fully take account of the imponderable problems of substitution. For certain industrial uses aluminium *could* replace steel. In the construction industry the amount could be reduced by the greater use of concrete, albeit much of it of the reinforced kind which contains steel strengthening rods. Nor do these estimates completely acknowledge the development work aimed at producing a more fuel efficient car by replacing the heavy cast-iron exhaust manifolds with heat resistant plastics as well as other ancillary items related to the power unit and the use of moulded plastic sections to replace steel in the bodywork. Translating comments made at the beginning on the abundance of iron ore into some quantitative measure, the USBM suggests world reserves stood in 1980 at 103,000 million tonnes inside a probable resource base of 217,000 million tonnes. Reserves are well distributed through the world and though the largest share of these is held by the USSR (30%), the rest are evenly spread through eleven key countries (plus some others) covering all the major continents. For these reasons the availability of supplies seem assured in the short and medium term.

## 6.2 Non-ferrous metallics

### 6.2.1 Manganese, chromite, nickel, cobalt, molybdenum, tungstèn, vanadium

The common factor which links all these metals together, the 1980 outputs of which are shown in Figs 6.3–6.5, is that their industrial usage is largely, but not entirely, connected with that of steel production. Whilst they all have an important role to play particularly in the making of specialized steels, they represent a group of metals across which substitution is largely possible even if at additional cost. When alloyed with steel all of them impart the qualities of strength, toughness and hardness. It is in this role that they find 80 per cent or so of their market, the rest being consumed in the chemical industries in the making of such products as dyes, pigments, catalysts and so on. If any distinction may be made here between these minerals it is that as an alloy cobalt is perhaps supreme particularly where very high strength and durability at high temperatures is demanded. That it is not more widely used than the other metals is due to its greater cost (resulting from its greater rarity and its availability only as a by product of copper). It is almost three times the price of molybdenum which is also recovered with copper, but is mainly won from primary exploitation, and six times that of vanadium, another metal which is only available as a result of working other minerals (that is phosphate, titaniferous magnetite and uranium). Only where resistance to corrosion and oxidation is a desirable characteristic of an alloy may nickel be said to have intrinsic merits over and above its competitors and these are not related to price.

For hard facing alloys used on tools, tungsten is a peculiarly suited and much less expensive substitute for molybdenum. Chromite may be distinguished by its unique position in the production of stainless steel and in the plating of other metallic surfaces. Grades suitable for such metallurgical uses (as opposed to those for chemical and refractory purposes) once needed to be of high chrome content (over 46%) and low in iron; now ores less rich in chrome are acceptable. However, one of these metallics, manganese, functions in a way not shared by the rest. When mixed with steel it has a desulphurising, deoxidising and conditioning effect upon it. For this reason its utilisation by the metallurgical industry rises to around 95 per cent of total output.

Manganese deposits show great variety of geological type although they appear most infrequently as oxides. The USSR deposits in the Nikopol Basin of the Ukraine and Chiatura in Georgia, which in 1980 produced nearly a half of world output, are in the form of flat lying or gently dipping sedimentaries of relatively recent geological age. In contrast the Indian deposits of Madhya Pradesh are found in the form of tabular lenses conforming with the surrounding intensely contorted and metamorphosed ancient sediments though weathering has given rise to important residual deposits. In the form of manganese carbonate (or rhodochrosite) this mineral can also be ident-

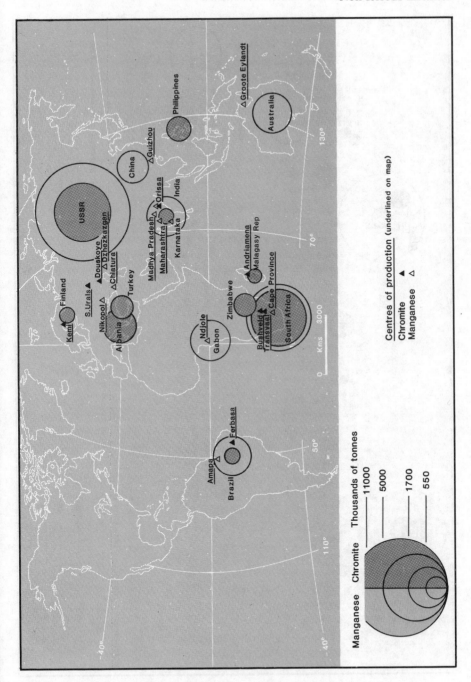

*Fig. 6.3* Manganese, chromite: output by country with chief centres of production, 1980. *Source: Mining Annual Review, Minerals Year Book*

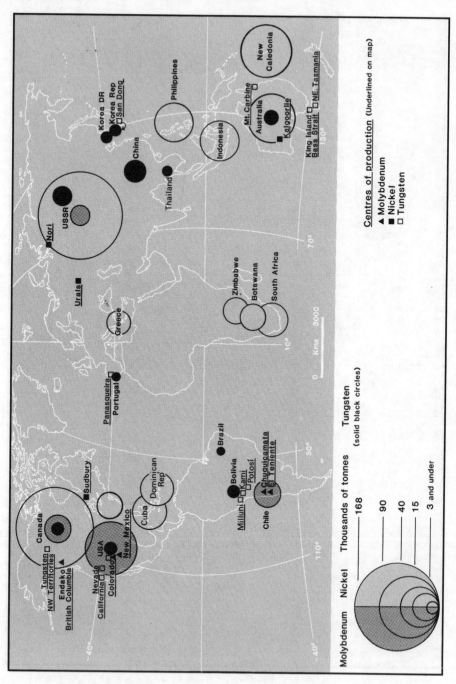

*Fig. 6.4* Nickel, molybdenum, tungsten: output by country with chief centres of production, 1980. *Source: Mining Annual Review, Minerals Year Book*

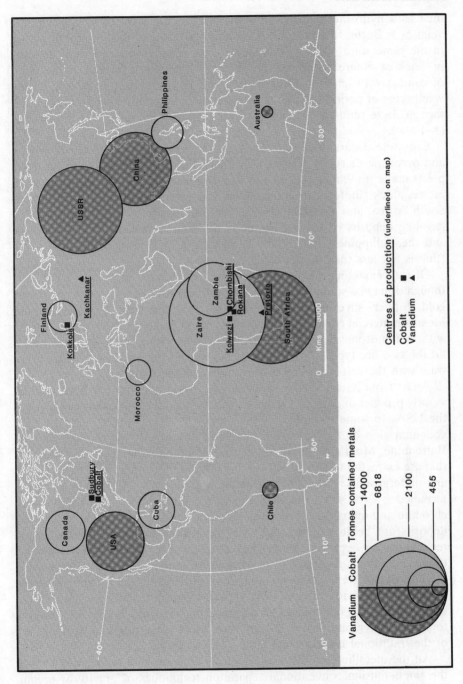

*Fig. 6.5* Vanadium, cobalt: output by country with chief centres of production, 1980. *Source: Mining Annual Review, Minerals Year Book*

ified as a hydrothermal deposit sometimes in the form of steeply dipping veins as at Butte, Montana, USA, or as a replacement body as at Philipsburg in the same state. Ores of this metallic are mainly obtained from deposits in veins or fissure fillings of nickel–copper sulphides formed at the point of contact with igneous intrusive materials or from laterites formed by the weathering of peridotite, dunite, pyroxenite or serpentine over a long period though these tend to be richer in iron (between 30 and 55%) than nickel (1–1.5%).

Chromite occurs in certain types of utramafic rocks composed of olivine and pyroxene or rocks derived from them and in two forms. Stratiform deposits make up 98 per cent of the world's resources and are usually several metres thick, uniform in composition and extensive (as in the Transvaal, South Africa, and also Zimbabwe). The rest consist, as with manganese, of lens-like deposits of varying extent to be found, for example, in the Urals and the Philippines, but usually containing a high level of chromic oxide (that is not less than 46 per cent).

Of the remaining metals only tungsten is extracted in its own right even though it may be sometimes associated with deposits of tin, antimony, silver-gold, copper, silver-copper-lead-zinc and, most significantly, molybdenum. Most commercial occurrences have been found in quartz veins and in contact with metamorphic deposits and are associated with siliceous granite rocks. Of the two ore types, 75 per cent of the world's deposits are of calcium nitrate with the rest in the form of iron-manganese tungstates.

Apart from its association with tungsten, molybdenum has been worked as a by-product of copper, particularly in the country of greatest production, the USA. In some instances, as the copper has been worked out, the molybdenum side of the operation has been developed (for example, at the Butte mine, Montana, USA). In the case of all the minerals referred to here that are extracted independently of any other mining operation, both surface and underground methods of extraction are employed, the mode being determined by the size, shape, depth and regularity of the deposit. With chromite and manganese open pit operations tend to prevail where lens-like deposits are being worked, but with the stratiform deposits underground techniques are deployed. For nickel the main sulphide ore sources such as those at Sudbury in Canada are worked by underground methods leaving the laterites of New Caledonia and Cuba to be exploited by open-pit techniques, a mode with only minority appeal. Open-pit techniques are also unusual where tungsten is concerned since lens-like ore bodies of the kind found near Plymouth in the UK are rare. More often tungsten workings are of the traditional underground kind.

All the metallic ores discussed here are concentrated before despatch to the smelter using conventional separation techniques of gravity or a combination of flotation and/or magnetic methods. Only for nickel obtained from laterite deposits has it so far proved impossible to undertake any form of concentration.

The overall demand for all these minerals is inevitably affected by the demands made upon the steel industry as a whole and thus upon the economic health of individual producer and consumer nations. However in referring back to what was said in the earlier section on iron and steel, certain caveats need to be entertained. As manganese is a vital input to steel-making, demand will bear a direct relationship to the output of steel though trends of the early 1980s suggest that the consumption of manganese per tonne will fall primarily as a result of steel-makers moving to more efficient processes, especially the greater use of electric arc furnaces. These effects are most likely to be felt in those countries in the forefront of such developments, particularly Japan and the German Federal Republic as well as the USA. Ultimately with other steel producing countries following suit, by the end of the century consumption per tonne of crude steel may well have fallen from about 6.6 kg per tonne to 6.1 kg per tonne.

As for cobalt the current users of alloys containing this mineral, those making critical rotating parts of jet engines which constitute the major part of the market, are concerned about its high cost and uncertainty about its future supply since 70 per cent is obtained from Central Africa. They are looking for alternative low or no cobalt alloys. Some progress has been made by replacing some of the cobalt in these alloys with an addition of nickel. As for its use in the chemical industry, it is rapidly being replaced where cheaper substitutes are available. Only in catalyst applications has it not suffered substitution.

Regarding molybdenum, the picture for producers has been brighter with demand rising and running against the economic trend of the early 1980s. Free from the supply constraints of cobalt, molybdenum is able to meet the needs of an oil industry intent on the development of wells in much harsher environments than hitherto. High-strength low alloy steels containing molybdenum are the common materials for the construction of pipelines where good weldability, toughness at sub-zero temperatures and high strength in heavy-walled large diameter pipes are needed. At the same time the desire to replace oil-based energy with that of alternative sources has created considerable expansion in the construction of solar panels. By far the most widely accepted material for such panels is a ferrite stainless steel containing 2 per cent molybdenum. High molybdenum content low alloy steels are also required in the manufacture of the increasingly popular front-wheel-drive cars, whilst greater defence requirements are bound to demand more aero-engines which are both quieter and fuel efficient and thus the greater use of steel/molybdenum alloys as well as those containing vanadium.

Finally, rising environmental standards will continue to require a reduction in $SO_2$ in the atmosphere (see Ch. 8). Scrubbing equipment suitable for the work will contain stainless steels with up to 3.5 per cent molybdenum where it is in general use, but for particular situations up to 16 per cent.

The forecast rate of annual increase in demand for molybdenum to the end of the century is, according to the USBM, 4.5 per cent. With reserves of 9.5 million tonnes and a resource base speculatively 20 per cent greater than that amount, both well distributed through the continents except Australasia (though North America is best favoured), future supplies should pose no problems.

Just as iron ore producer countries were observed in the previous section to be moving into the business of steel production, so the alloying of metals is increasingly moving to the countries producing the ores and away from the traditional steel manufacturing countries. This must further enhance the capability of the countries producing the ores for alloying purposes to add value to their products. South Africa, the leading producer of chromite, had taken its share of the world production of chrome alloy from negligible figures in the early 1970s to over 50 per cent by 1978 and over 60 per cent by 1980, though it has to be recognised that this country has the additional advantage of low energy costs through easily won coal located close by the sources of ore. Six new chrome alloy plants were established in the ore producing area early in the 1980's, all geared to the export market.

As a consequence of the trends illustrated there has been a marked reduction in the volume of, for example, imports of manganese ore, chromite etc. by most of the major steel producing countries and a sharp increase in manganese and chrome alloy imports, a situation which can be expected to continue. With regard to a number of these metals whose demand is inextricably tied to that of the steel industry, the questions of future supplies may seem less of a problem than at any time in the last decade. For the most essential of all the minerals, that of manganese, the USBM maintain that the world's reserves are in excess of 1500 million tonnes of contained manganese out of a probable resource base of double that amount. Some 53 per cent of reserves are available in South Africa and 26 per cent in the USSR. The rest of the current reserve is found primarily in Australia (9%), Gabon (3%) and Brazil (3%). Where nickel is concerned, known reserves run at about 60 million tonnes out of a resource base of possibly 228 million tonnes, but the estimates made by the *Mining Annual Review* are essentially based on incomplete information and reserves may well be in excess of that figure. As compared with manganese, the mineral is found in significant quantities in at least ten countries outside China and COMECON, with Canada having by far and away the greatest potential. Though not quite so widespread, cobalt reserves are found primarily in Africa (established at 1641 million kgs), Australia, New Caledonia and the Philippines (318 million kgs), the USSR (227 million kgs) and North America, including Cuba (209 million kgs), with the chief non-Communist producers of the sulphide deposits being Zaire, Zambia and Canada. The exploitation of this mineral is almost entirely dependent on the extraction of its associate metals, nickel, copper, vanadium, chromite, lead, zinc, uranium and manganese.

Of the above three metals, the fact that the manganese requirement per tonne of steel produced is being reduced might seem to assist the supply situation even further. However such optimism is not justified for manganese since there could be medium-term problems. Several factors appear to contribute to such a possible situation. First, in less than a decade the COMECON countries which were net exporters (primarily Russia) have become net importers. Moreover, the published Russian 'reserve' figures are considered by some to be highly questionable with high grade reserves close to depletion and many of the typical grades of ore too low to fall into this category by world standards. Second, although the South African reserves are the biggest in the world, grades are declining steadily, costs rising and no new ore bodies have been discovered there for twenty years. Yet the free world might have to become dependent on this country by the year 2000. Third, the Australian and Gabonese ore bodies are clearly defined and depletion may be anticipated early in the twenty-first century. Fourth, there are no known land-based sources which *could* make up anticipated deficits amongst current producers. Finally, erstwhile exporters of ore (though relatively the only marginally significant ones) seem likely to become significant importers in the next decade (that is India, Brazil and Mexico).

This state of affairs has to be seen against USBM forecasts of demand. That organisation suggests an average annual growth in the manganese requirement of 2.7 per cent to the end of the century. As it happens where manganese is concerned – and it is also true of nickel and cobalt whose annual demand rates are predicted to grow by 4.3 per cent and 3 per cent respectively – the supply situation *could* be radically changed for the better in the not too distant future as these minerals have been identified in nodules on the floors of the Pacific, Atlantic and Indian Oceans and currently play no part in reserve estimates. So far it has been suggested that about 1.5 trillion tonnes of such nodules, containing 24 per cent manganese, 1.6 per cent nickel, 0.21 per cent cobalt and 0.05 per cent molybdenum (as well as 1.2 per cent copper), lie between depths of 30 and 3,000 metres, though estimates of the richness of the nodules have varied (Enzer 1980). The success of their recovery will depend on the provision of a system capable of hoisting between 1 and 3 million tonnes (dry weight) of nodules from the sea bed each year and determining by international agreement the division of such a rich harvest between countries. Both are within sight of a solution, the effect of which would be to widely extend to a number of nations self-sufficiency in these minerals. However, it must not be forgotten that the leading land-based producer nations of the minerals in question (especially Zambia, Zaire, Zimbabwe, and Canada) will also be affected but adversely by ocean mining. For this reason they have sought to curb the immediate impact of sea-bed nodules by an international agreement to be ratified in 1988. This will permit only world *growth* in demand for these minerals to be obtained from nodules during the first five years of their production. Thereafter deep sea mining will not take more than 60 per cent of that growth and land based production not less than 40 per cent (United Nations Law of the Sea Conference, Geneva, September 1980).

Of the three remaining minerals, vanadium, which is not mined in its own right, is heavily concentrated in one country, South Africa, with over 90 per cent of the known reserves of about 15,818 million kgs and 70 per cent of the world export market. Both world super-powers and China largely share the remainder of the reserves. In spite of the recession of the late 1970s and early 1980s, USBM forecasts an average 5 per cent per annum growth in world demand up to the end of the century due to an increasing demand for high performance materials in which weight saving is an important element (as in the transport industry).

Tungsten is considered to have reserves of 2,590 million kgs of contained metal out of a possible resource base of almost 6,818 million kgs. China is the best placed with around 53 per cent of the world reserve total. The rest is well distributed through North and South America and Europe (including the USSR). Assuming that demand increases at around 3.5 per cent a year and allowing for some growth in the rate of recovery of tungsten from both manufacturing processes and recycling to around 3 per cent of total demand by the year 2,000, no shortages of the material are likely to occur.

Chromite, last of the metals related to steel production, is probably the most problematic. It plays a unique role not only in plating, but also in the production of stainless steel where it is vital. World resources have been identified at about 33,000 million tonnes, ten times the reserve figure. With a likely demand rate of increase per year of around 3.3 per cent to the end of the century on a 1980 production figure of 9.9 million tonnes, no problem in terms of world supply could be envisaged. However, the USSR has access to considerable quantities of this mineral, much of it of the high grade metallurgical kind, and is probably the second largest producer immediately behind South Africa, although there are no exact figures, nor can it be established with any certainty that it has a long term reserve potential. At the same time 98 per cent of world reserves outside the communist bloc and 99 per cent of known resources come from South Africa and Zimbabwe, both countries which for different political reason may be considered potentially unstable. It is small wonder that of all the minerals that the USA attempts to stockpile against interruption to supply during war, it considers chromite amongst its most important. Both the USA and Japan have anxiously attempted to find substitutes for its use wherever possible and to find alternative supplies by investigating potential deposits in Brazil, Papua New Guinea, Indonesia and the Philippines.

## 6.2.2 Aluminium; magnesium; titanium

*Aluminium*

On two counts aluminium may be considered to be a metallic of major importance. First, in its basic form (it is mined as bauxite which consists of a mixture of hydroxides of aluminium) it is the most abundant metallic substance in the earth's crust (8.3%). Second, its utilisation in practically all sectors of the economy has shown a faster growth rate than that of any other metal. Whether measured in quantity or value, the amount of aluminium consumption is exceeded only by that of iron. The basic reason behind this has been its ability to substitute for other less easily won or rather more scarce minerals and its lightness and resistance to corrosion. As for its former quality, its replacement for copper in transmission cables is of considerable significance. Since it conducts heat almost as well as it conducts electricity it has replaced steel and copper in many kitchen utensils. Its lightness and corrosion resistance make it an obvious choice for the construction of boats and vehicle bodies where it can also substitute for steel and, to a lesser extent, wood. Aluminium is also used in a mixture with small quantities of copper, manganese and magnesium to make a metal which is as strong as steel but very lightweight, Duralamin. This is an invaluable material in the aircraft industry but is expensive to produce.

Bauxite is formed by the weathering of aluminium bearing rocks under conditions favourable to the chemical breakdown of minerals in the parent rocks, the retention of the aluminium as hydrated aluminium oxides and the leaching out of other constituent materials. Since it is a soft earthy substance, bauxite is easily dug and requires no drilling or blasting. About 90 per cent of output is produced using open-pit techniques. Over-burden is relatively thin (the thickest coverings which occur in Surinam and Guyana are only 60 metres) and it is frequently returned to the site once the bauxite is extracted. The cost of extracting the raw material represents some 10 per cent of total costs of producing the refined metal compared with 80 per cent for copper. That this situation results from something more than the ease of extraction of bauxite is a matter for further discussion later.

Although bauxite is found and worked in eighteen countries (outside the USSR, its satellites and China), the large producers are mainly those of the less developed world (chiefly the tropics) and are spatially removed from the main centres of consumption. Of the five nations which produce about 78 per cent of the world's production, the largest producer (and the one exception to the comment about less developed countries) is Australia. It produces around 36 per cent of the world's bauxite, followed by Guinea (16%), Jamaica (16%), Surinam (6%) and Guyana (4%) (Fig. 6.6). In contrast the major consumers are the industrialised nations of the world. Although in 1980 some forty of these produced aluminium, the USA led with 29 per cent of the world's output of refined metal, the USSR 16 per cent, Japan and Canada 7 per cent each, the German Federal Republic and Norway under 5 per cent each. Amongst the major primary metal producers, only France and the USSR were able to produce the greater part of their bauxite requirements.

Apart from the usual socio-economic factors, demand forecasts for aluminium inevitably focus on its continuing role as a substitute material, particularly the likely increases in its use due to growth of electricity as a main

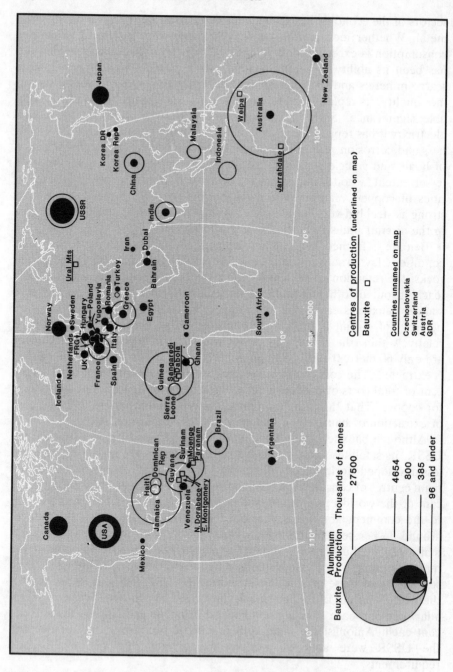

*Fig. 6.6* Bauxite, aluminium: output by country with chief centres of production, 1980. *Source: Mining Annual Review, Minerals Year Book*

source of power. Here the less developed countries are likely to be of special significance in so far as demand is concerned as their power distribution systems are as yet relatively rudimentary.

Taking all these factors into account the USBM postulated an average annual growth rate of demand up to the end of the century of about 5.4 per cent. It also estimates that bauxite reserves are running at over 22,000 million tonnes, a quarter of which would represent the reserve of recoverable aluminium equivalent. Thus with the strong likelihood of further discoveries of workable bauxite and the fact that there are other sources of aluminium besides bauxite (for example, clays of the kaolin type, anorthosite and alunite), supplies for the future would seem entirely adequate.

So far in the disccussion of metallic ores the relationship between sources of supply of the raw materials and the areas of demand has been noted as well as spatial changes in the intermediate processes of beneficiation and the conversion of the ores to metals. Aluminium offers a slighly different picture because of the particular processes through which the raw material passes before conversion to metal. At the point at which bauxite is extracted, crushing, washing and drying operations are undertaken but the material does not require beneficiation.

The process of drying sufficiently obviates the shipment of both unnecessary weight and bulk. However a further process is required in advance of the production of the metal since the bauxite is given a hydrochemical treatment to extract hydrated alumina oxide from which alumina is produced. The alumina is then subjected to a process of fused salt electrolysis in a molten bath of fluorine salts to extract the aluminium metal.

The points at which these various processes are undertaken have changed markedly. As the production of aluminium metal expanded after 1905 (following the invention of the relevant extractive technology), the bauxite mines themselves were only engaged in the rudimentary first stages of processing the material prior to its despatch to the industrial countries for conversion to alumina and then to aluminium. More recently the trend has been for bauxite producer countries to produce alumina with the relevant plant closely related to bauxite mines. However, the final stage of the conversion to aluminium suffers from one major disadvantage – the large amount of power consumed per kg of metal produced by the electrolysis process. In the early 1980s this averaged around 3.6 kWh. This makes smelting costs high for aluminium (65–70% of total metal production costs compared with only 20% for copper).

Following the massive rises in energy costs in the middle to late 1970s, many of the smelting plants of Europe and Japan which were heavily reliant on imported oil, found themselves uneconomic. The result is the emergence of a new spatial pattern in which the final metal processors are being located either where there is indigenous bauxite plus cheap energy or where there is inexpensive energy alone (Fig. 6.6).

Australia, the major producer of bauxite, planned a capacity increase for the production of primary aluminium from 370,000 tonnes per year (1980) to 1.3 million tonnes per year (1985). The key factor here is that as well as having the world's largest bauxite reserves (4,600 million tonnes) it has huge reserves of easily mined coal in the vicinity from which very cheap electricity may be generated. Given the highly stable political background of the country, investment capital is not a problem.

Guinea, with over 15 per cent of the western world's current output of bauxite and with some smelter capacity already available, has a major bauxite–alumina–aluminium complex planned at Konkoure (700,000 tonnes per year) to be based on a hydro-electric power scheme. Similar plans are being made by Guyana where its first smelter, with a capacity of 150,000 tonnes per year, will be powered by a plant on the Upper Mazuruni River. Another bauxite producer which plans to construct smelter capacity is Indonesia with a 225,000 tonnes per year project powered by the Asahou River in the Kuala Tanjuing district of North Sumatra. Brazil already has some 250,000 tonnes per year of aluminium smelting capacity but proposes to expand this with a 160,000 tonnes per year complex based on power from the Tocantins River in Moranhao State.

However, India's aluminium smelting industry, based on its own reserves, suffered a 15 per cent fall in output between 1979 and 1980 due to its inability to afford oil imports. It expects to resolve this problem by building a 218,000 tonnes a year smelter based on local cheap coal.

Finally, where the bauxite producers are concerned, Mexico, a country with plentiful supplies of its own oil expects to double its present aluminium smelting capacity to 90,000 tonnes per year. Venezuela, in much the same position regarding the availability of indigenous hydrocarbons, plans to double its present capacity to 610,000 tonnes per year by 1985.

Several other countries not associated with the mining of bauxite intend to increase their capacity to smelt imported ore using hydro-electricity. Canada (a country which in 1980 produced 8.4% of the non-communist world's aluminium from US bauxite imports) intends to expand production capacity in British Columbia and Quebec. The UK (producing 3% of the world's aluminium outside COMECON and China) raised its Lochaber capacity to 37,000 tonnes per year in the early 1980s. Both Argentina and the Philippines intend to base their smelters, in the 135–140,000 tonnes per year range, on Australian alumina and bauxite respectively using local hydro-electric power.

Aluminium smelting capacity has been hardest hit in Japan by rising energy costs mainly because it was based almost entirely on oil. In 1979 500,000 tonnes of capacity was shut down and there is little prospect of any home-based new capacity being developed in the foreseeable future. Japanese consumers are therefore pinning their hopes on taking financial stakes in some of the smelter developments described previously in Venezuela, Australia and Indonesia. The last named is expected to export two-thirds of its output of aluminium metal to Japan.

However, it should not be assumed that the demands of aluminium production in terms of high levels of energy consumption and cost are immutably and irrevocably changing the spatial patterns of bauxite mining, alumina production and aluminium smelting. Experiments aimed at reducing the energy input to smelters have been successfully put into practice in South Carolina, USA, where a new 180,000 tonnes per year capacity smelter uses only 2.8 kWh of electricity per kg of aluminium metal produced instead of the common average of 3.6 kWh per kg.

Another check on spatial change could be the much greater use of recycled aluminium. In the USA about 80 per cent of the available scrap is currently being lost because of a basic inability to sort it from other forms of refuse. However, the incentive to discover adequate ways of undertaking this must be a significant one for developed countries without indigenous supplies of bauxite and an energy base which is largely dependent on imported oil because the recycling of aluminium metal requires only 5 per cent of the energy input involved in the production of primary metal from alumina.

*Magnesium*
Magnesium in its metallic form, like aluminium, is light and expensive to produce if the technology currently in greatest use is deployed. This electrolyte process demands high levels of energy input of 3.6–4 kWh per kg of metal fabricated. That this situation has not had a significant influence on the location of smelting plant favouring countries with low cost electric power is due to the fact that in recent years a new silicothermic process, which uses much less energy and has the additional advantage of producing only low levels of environmental pollution, has come into use. In 1980 alone all major producers of magnesium metal in the western world were substantially adding to their silicothermic capacity at the expense of electrolytic processing.

Although magnesium metal has an important role, as has been seen in the making of Duralamin, and is of great importance in the production of alloys for air-cooled car engines, it is also widely used in oxide form in the manufacture of metallurgical furnace refractories, as a vulcanising agent in rubber, and for other miscellaneous chemical products. The refractory aspect of its use has become of growing importance in recent years where its capacity to provide a hot enclosure (or lining) in which certain industrial processes may be conducted has been increasingly recognised. This is especially so in the steel industry which has undergone rapid technological change. Thus whilst other forms of refractory, such as the many forms of silica rock, were important to the open hearth steel making process, since the 1930s the newer basic oxygen and electric arc methods of steel making have come to the fore because of their greater efficiency and economies in the use of fuel. These have demanded a refractory such as magnesia in one of its high quality forms mixed with small quantities of chrome. Such magnesia-based refractories can withstand the more testing operating conditions both of temperature and of chemical and physical attack. Similar trends have been detectable in the manufacture of cement. As kiln size has increased in capacity basic refractory linings using a magnesite base have been favoured. These now make up almost 50 per cent of the lining material used. There has been a general acceptance that although such refractories command high prices, these are more than justified by their longer lives, their higher standards of performance and lower maintenance requirements.

The sources of magnesium for whatever use distinguish it from all other minerals so far considered. Not only is it the third most abundant structural element in the earth's crust, but its supply is virtually unlimited. It may be extracted from the relatively ubiquitous magnesite and dolomite rocks as well as from the sea where it is estimated that every cubic km contains a million tonnes of magnesium. It follows that, although countries with excellent supplies of, for example, dolomite, may be immediately favoured (such as Yugoslavia) most of the developed countries of the world are in a position to have access to magnesium for the foreseeable future. Where the source of magnesium is dolomite or magnesite, the raw material is extracted (usually from open pits) in the normal way and processed as if it were an aggregate before being reduced to fine powder when the mineral in question is removed by flotation. When the source is sea water (or water from inland salt lakes) magnesium lime is added to precipitate out magnesium hydroxide. This, in the form of a slurry, is filtered and then calcined. 1980 production by country is shown in Fig. 6.7.

The future demand for the material, largely unconstrained by the supply situation, is dependent on the usual socio-economic factors plus substitution possibilities. In its metallic form aluminium and zinc can be substituted in many die casting applications and as a refractory it can be replaced by alumina and chromite. Most of the evidence suggests that magnesium itself will substitute for the materials mentioned rather than the reverse and USBM has estimated an annual growth rate in demand of 2.1 per cent to the end of the century.

*Titanium*

In its metallic form titanium is used as an alloy where its strength and capacity to resist corrosion even at very high temperatures are much valued. It thus appears in the manufacture of ships, aircraft and space vehicles where sections most likely to be subject to intense heat are made from 85 per cent titanium alloys. Titanium also finds its way into paints and varnishes where as titanium dioxide it imparts whiteness, opacity and brightness. These dioxides are not expensive to prepare but they share with aluminium and magnesium in the production of their metallic forms a propensity to be a very high cost commodity. The production of sponge metal (that is the stage before final conversion to ingots) can involve an energy consumption of 3.2–10 kilowatt hours per kg depending on the nature of the processing plus a fur-

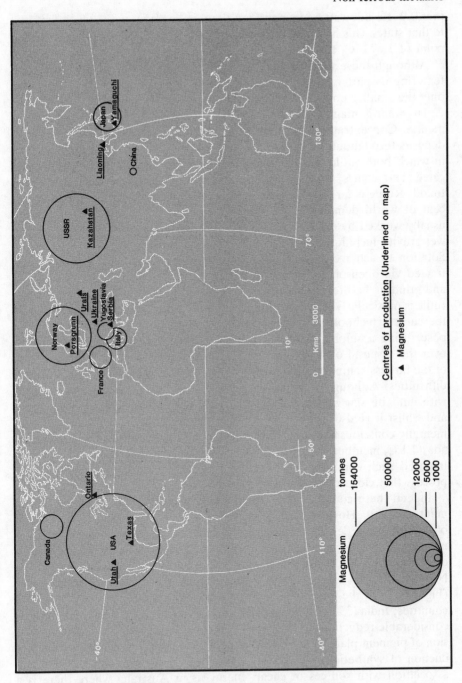

*Fig. 6.7* Magnesium: output by country with chief centres of production, 1980.
*Source: Mining Annual Review, Minerals Year Book*

ther 0.9–1.1 kilowatt hours per kg of titanium ingot to convert sponge metal to that state. This is hardly surprising when it is realised that it has a melting point of 1,675 °C, the highest of any metal.

Although these metallic uses are highly important and result in titanium featuring as part of the US stockpile of strategic minerals, they represent only the smaller end of the market for the mineral.

Titanium is mainly derived from deposits of two minerals, rutile and il-menite. Concentrations of these occur only in a relatively small number of deposits throughout the world. They are either placer concentrations of sand in which both rutile and ilmenite occur together and are almost always lo-cated near coasts, or they are rock deposits where ilmenite is invariably found. Rutile is far less common than ilmenite which supplies around 88 per cent of world demand for titaniferous materials. The placer deposits are usually worked by dredge and the sand is separated from the minerals using wet gravity techniques. The valuable mineral is often further treated using flotation to achieve greater concentration. Ilmenite in rock deposits is ex-tracted via open-pit techniques followed by the usual preliminary crushing and grinding before concentration. The balance of supplies of ilmenite and rutile is particularly significant in their major role as pigments. Unfortunately the current technology for converting the commoner mineral for this pur-pose, using a sulphate process, can cause severe environmental problems over the disposal of waste acid and ferrous sulphate, whilst the conversion of the far less common rutile using a chloride process presents far fewer such difficulties. Although a compromise process involving the blending of rutile with ilmenite slag which is conducive to lower effluent outputs does exist, and whilst it is also possible to produce synthetic rutile, the more environ-mentally conscious countries of the developed world such as the USA, Can-ada, EEC member states and other Western European countries, have imposed such conditions on pigment producing plants using the sulphate process that closures of plant have been rife.

Indeed these countries have traditionally been the major centres of pig-ment capacity. However the propensity for such processing to be pushed back to the less developed countries, which are frequently not so environ-mentally conscious producers of raw materials, does not exist since the prin-cipal ilmenite sources are Australia (33% of world output), Canada (23%), Norway (15%), USA (13%), South Africa (10%) and Finland (3%) (Fig. 6.8). Of the other major producers only three are less developed countries; India (3%), Malaysia (3%) and Sri Lanka (2%). The result must be considerable reductions in the use of ilmenite as a pigment, the rapid conver-sion of pigment plant to less environmentally damaging processes, or the pro-duction of synthetic rutile, although the latter really only appears likely in association with sources of cheap energy as in Australia where there is a close coincidence of coal and ilmenite supplies. Self-evidently in terms of the use of ilmenite, whether as a metal or as a pigment, many of the producers of the mineral are also consumers. However, international trading relations emphasise the pre-eminence of certain major manufacturing nations in demand terms. Taking the major producer, Australia, 33 per cent of its exports go to the USA, 17 per cent to the UK and 6 per cent to Japan. The three less developed countries are also exporters to these and other developed countries.

Where rutile is concerned the pre-eminence of Australia as a producer is even more obvious with about 90 per cent of world output coming from that country. Other and relatively minor sources of supply are South Africa, the USA, Sri Lanka, India and Sierra Leone with the export potential of the last named developing rapidly along with that of South Africa, China and the USSR (Fig. 6.8). Again the USA was the largest importer of Australian titanium, this time rutile (41% of its exports), followed by the UK (17%) and Japan and the Netherlands (11% each).

With both types of titanium the countries working the resource are all involved at least in the preparation of concentrates. However, whatever the end use, Australia is far and away the most important producer of concentrates alone, though it would follow that in terms of added value, some capacity to produce sponge is both logical and financially sound. Other smaller producers including India have certainly become or are about to become involved in the export of sponge metal. China has emerged as a regular and increasing seller of sponge to the USA, along with the USSR though the latter on a much smaller scale. Other developed manufacturing countries without access to the basic minerals but which are importers of concentrates of them have also, unusually, become involved in an intermediate metal fabricating stage designed not merely to meet home needs but also for export. Japan is a prime example with a fast growing export market for sponge.

The USBM calculated that world reserves of rutile for 1983 were about 81 million tonnes of contained mineral out of a likely resource total of nearly 170 million tonnes. As for ilmenite the resource figure is now over 600 million tonnes of which 220 million tonnes may be classed as reserves. The usual socio-economic factors affecting the rise in the gross domestic product of both the developed and the less developed countries will likely influence future demand for titanium. Although substitution may occur where titanium is used to achieve corrosion resistance (for example, stainless steel), for aero-space applications (for example, aluminium alloys) and in its utilisation as a pigment (for example, zinc oxide, talc, alumina), all the replacements are either not as cost-effective or are qualitatively less successful. The USBM therefore point to an annual growth factor up to the end of the century of about 3.8 per cent.

The predominance of one supplier, Australia, which is both stable and democratic should mean that no supply difficulties will arise as far as the other free market economies are concerned, though Australia has used the withdrawal of titanium exports to the USSR as a political weapon. Perhaps of more importance in world trade will be the fact that large supplies of Australian titanium are contained in placer deposits the working of which,

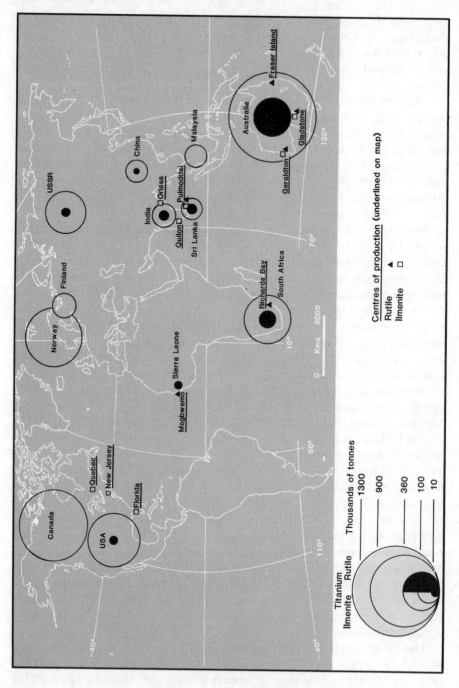

*Fig. 6.8* Titanium: output by country with chief centres of production, 1980.
*Source: Mining Annual Review, Minerals Year Book*

by dredger, can cause major landscape despoilation. This situation has already led to a protracted dispute over a large potential mining operation on Frazer Island between a mining company and environmentalists.

### 6.2.3 Copper, tin, lead and zinc

These metals are grouped together not only because of their considerable importance as key industrial minerals and as internationally traded commodities, but also because of their long history as resources valued by man. Copper has probably been in use for at least 6,000 years, first used in an unalloyed form for tools, weapons and ornaments, and subsequently, by the addition of tin, as bronze, giving its name to an era of cultural development that diffused across Europe from the Near East. Similarly, an alloy of copper and zinc – brass – has been employed extensively for over 2,000 years, probably having first been used by the Romans. As for lead, it is frequently mentioned in the Old Testament. Lead pipes occur in constructions dating from dynastic Egypt whilst the hanging gardens of Babylon were floored with lead sheeting. It was also used at that time to glaze pottery and in the construction of ornaments.

Within this broad grouping, however, certain distinctions exist in terms of the spatial aspects of supply and demand for these minerals and their geological origins and associations. Copper and tin are major materials in the trading dialogue between developed countries and less developed countries. Of the ten major producers of tin outside the USSR and China, seven are less developed countries contributing 84 per cent of the total output of primary metal concentrates (Fig. 6.9). On the other hand 71 per cent of primary tin is consumed by only nine major developed countries with the two key users, the USA and Japan, taking 25 per cent and 17 per respectively. As the economies of some producers begin to expand a limited amount of internal consumption is becoming apparent. The new industries of Brazil, for example, increased their consumption of tin by 20 per cent from 1979–1980. However, it does remain largely true that any attempts to add value to the mine output earnings by the less developed countries have largely been confined to their smelting the ore concentrates before exporting the metal.

As for copper, the six major less developed countries exporting this mineral produce almost 50 per cent of total output of the non-communist world (Fig. 6.9). The economies of three of them (Zaire, Zambia and Chile) are largely dependent on the winning of copper for the earning of much needed foreign exchange. The chief industrial countries of the developed world consume 88 per cent of the available supplies of the metal including copper scrap which currently runs at about 40 per cent of annual total consumption. That the distinction between developed and less developed countries as consumers and producers respectively is not greater owes much to the 31 per cent contribution to non-communist world copper requirements by the USA

139

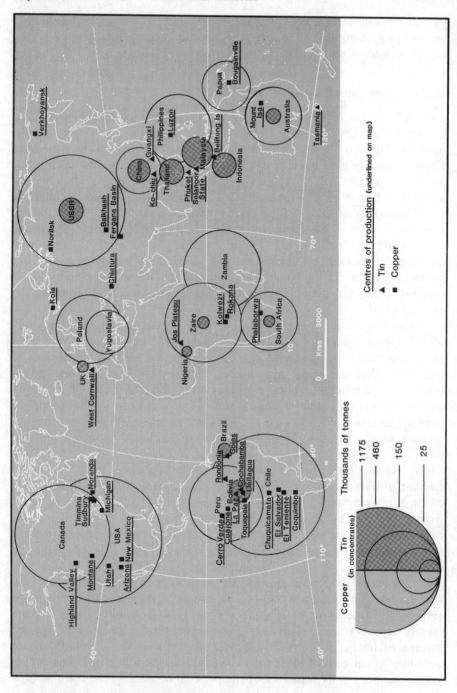

*Fig. 6.9* Copper, tin: output by country with chief centres of production, 1980.
*Source: Mining Annual Review, Minerals Year Book*

and Canada. As with tin, the major way in which less developed countries have increased their earnings is by a major shift away from the export of concentrates towards those of the metal itself. Zambia, for example, smelts almost all of its production whilst the figures for two of the other major producers, Chile and Peru, are 76 per cent and 61 per cent respectively.

For lead and zinc there is a much less clear-cut division between suppliers and consumers and developed, and less developed countries. Of the fifty or more countries in which lead is found, twenty-eight may be identified as producers of significance, with twelve of these each contributing over 100,000 tonnes per annum of metal content (Fig. 6.7). Outside China and North Korea, only three of these are less developed countries. Thus much of the available supplies are in the hands of the consumers. Moreover, an examination of those producing countries makes evident the wide spread of production sources throughout five continents. One particular distinction for lead is the very high contribution to world supplies of the metal by secondary sources (that is recycled lead) of some 40 per cent. Another is the capacity of many European countries with only modest mined outputs to perform as major suppliers of the metal (that is France, the German Federal Republic, UK, Italy and Spain), whilst one country, Belgium, which has no indigenous mining output, produces over 100,000 tonnes per annum of the metal. The USA output of refined metal which is the largest in the world runs at twice its mined average. It also leads the way in terms of consumption with 27 per cent of the world's annual total. The consumption of other major industrial countries in 1980 ran some way behind with Japan at 10 per cent, the German Federal Republic 9 per cent, the UK 8 per cent, Italy 7 per cent and France 6 per cent.

Because of its frequent geological association with lead in the same rock body, zinc shows similar patterns of supply. Of the fifty countries in which it is found, there are thirty-four with a significant level of extraction (Fig. 6.10). Fifteen of these each produce over 100,000 tonnes per year (metal content) of zinc although only two are less developed countries apart from China and North Korea. The potential for recyling zinc is much less than that for lead. Only some 20 per cent of world supplies are gained in this way. Nevertheless there are still major producers of the metal which do not extract the ore. Both Belgium (246,000 tonnes in 1980) and the Netherlands (169,000 tonnes) are significant with a number of other European countries which are major producers of refined metal but have only modest outputs of mined ore. The German Federal Republic, for example, smelts twice its own output and Italy well over that amount. Japan's home production of mined metal is only one-third of its smelted output. Thus unlike the other metallics under discussion, Europe and Japan offer major smelter outlets both for lead and zinc based on the import of ores. The reason for this may be connected with the wider 'spread' of the supply sources. With many of these sources associated with developed countries, the incentive to add value by both concentration and smelting is that much less. What may now tip the

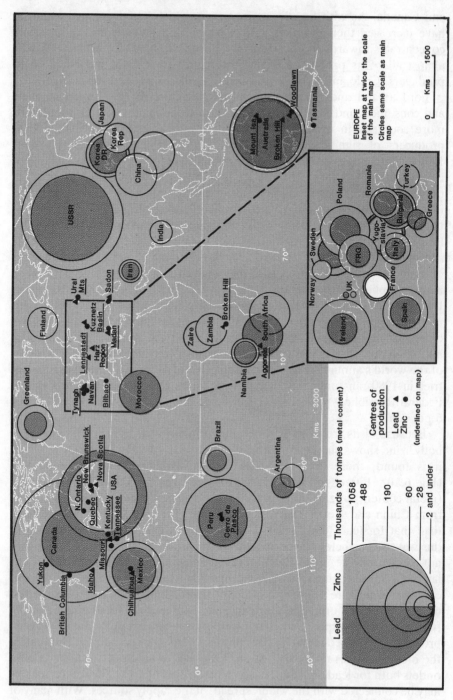

*Fig. 6.10*  Lead, zinc: output by country with chief centres of production, 1980.
*Source: Mining Annual Review, Minerals Year Book*

balance towards smelting in the producer countries could be the high energy costs in Europe and from the supplier's point of view, the opportunity to fully process ores which may not only contain zinc but small quantities of other associated high value metallics such as silver. Although much of the older smelting capacity in Europe lacks the facilities to smelt the higher value subsidiary ores, it is evident that Canada, with only about 43 per cent of its zinc ore output smelted at home, Mexico with 26 per cent and Peru with only 16 per cent, are all moving quickly in the direction of only exporting refined metals. However, the last two countries see themselves as well placed to supply other less developed countries with the refined metals once their economies begin to expand.

The geological association of zinc and lead is to be found in the two major types of deposit. Though vein deposits of these metals were the first to be exploited and a significant amount of production still emerges from such sources in Europe and South America, the major lead and zinc deposits are now the strata bound deposits or massive mixed sulphide deposits. With the former, commonly associated with limestones and dolomites and of generally lower grades, the metallurgy of separation from the parent rock body is simple and relatively low cost. The larger source of both lead and zinc remains, however, the latter type of deposit found in metamorphic rocks and irregular breccia. Here the percentage grades are higher and in spite of the fact that the sophisticated separation techniques involved make the extraction of the metals more expensive, the large sized deposits of this sort contain some of the great modern underground mines of the world. However, quite exceptionally there are a few open-pit operations in Canada.

The massive sulphide deposits also frequently contain copper along with lead and zinc, whilst the strata bound deposits are similarly favoured. In spite of the fact that some lead and zinc deposits contain relatively high grade copper ores (about 4% in the Zambia–Zaire copper belt) about 65 per cent of the world's copper resources are accounted for by porphyry deposits and vein and replacement deposits. Porphyry deposits are large relatively low grade bodies of disseminated copper (0.4–1.2% copper) associated with intrusions of felsic igneous rocks. Vein and replacement deposits are often associated with porphyry intrusions as localised concentrations in favourable host rocks. Important sources of low grade ore are to be found in Canada, Peru, Chile, Papua New Guinea and the Philippines.

Tin ore (casserite) is found in association with acid igneous rocks in the form of vein or disseminated replacement tin deposits. The main sources form a clear but irregularly demarcated belt surrounding the Pacific Ocean and along the eastern edge of the Atlantic and in south central Africa. However, the major sources of the metal ore (producing over 80% of metal ore output) are now no longer the originally worked vein deposits of areas such as Cornwall (UK), Australia and Bolivia, but the placer and alluvial deposits derived from the weathering and erosion of vein and disseminated replacement tin deposits. These are worked in Malaysia, Thailand, Indonesia and

Nigeria. Although the vein deposits are still worked by underground methods, unlike the other metallic ores discussed in this section, open-pit methods are widely used for placer and alluvial sources and the techniques deployed in the primary extraction phase are more akin to those for the winning of sand and gravel. Subsequently, the casserite is washed free from the parent material before concentration via conventional methods. Only for vein tin deposits is an additional rock crushing stage required between extracting the material from the parent rock and concentration.

The future demand for these metallic minerals will depend on the rate of growth in the economies of the less developed countries as well as on the continued expansion of the developed countries. Looking first at the known quantities of the ores it would seem that reserves of copper could be nearly 500 million tonnes out of a resource base of over three times that amount. In calculating the latter figure the USBM has taken into account surmised but as yet undiscovered deposits, plus those currently uneconomical and the reserve figure, but not copper in the form of nodules lying on the sea bed. These could add a further 690 million tonnes to the resource base by the 1990s when the technology for their recovery is likely to become available. Since the USBM forecasts a world annual growth rate of just under 4 per cent for this mineral to the end of the century, then by calculating cumulative demand year on year, the reserve base as determined in 1980 is adequate.

Tin reserves may be in the region of 10 million tonnes out of a possible resource base of 37 million tonnes. With demand forecast to rise at about 1.0 per cent a year according to the USBM (taking into account mined and secondary sources) by the end of the century, cumulative demand when measured against reserves indicates that there will be no supply problems. For both copper and tin it seems unlikely that any major changes in the supply pattern are likely to be forthcoming, though there may be some shift in the relative importance of particular countries. In respect of tin, Indonesia and Thailand could become the leading suppliers over Malaysia. As it is, Indonesia and Thailand have more reserves than Malaysia, and Thailand not only has more subeconomic resources than Malaysia but there is a strong possibility of further tin discoveries in that country.

The number of countries producing copper may be enlarged by the entry of Argentina and Columbia (from Andean deposits) and Brazil (sedimentary deposits). Porphyry deposits of Iran, Indonesia and the islands of the southwest Pacific could receive increasing attention. However, the major unknown will be the likelihood of procuring copper from the sea bed. This source could substantially add to the reserves of copper in the hands of the less developed countries currently standing at about 66 per cent of the total.

Lead reserves may be in the region of 127 million tonnes. With demand possibly running at a 3.5 per cent annual growth rate according to the USBM, the cumulative total of demand may approach or even exceed supply by the end of the century. However, already geographically widespread, the resource base is undergoing rapid expansion as a result of technological developments and renewed exploration activity. There is a high probability that mineable reserves will be augmented sufficiently to cover any reasonable forecast demand level. Australia, Canada and Yugoslavia, all currently significant producers, have in the early 1980s yielded what look like viable deposits and appear to be well set to supply between them total Japanese and European needs in the immediate future. Certainly recycling will need to remain an important and growing facet of the trade in lead.

Zinc reserves, estimated at around 162 million tonnes by the USBM will also be inadequate if its forecast of an annual growth rate in demand of 2.1 per cent for primary and secondary zinc is maintained to the end of the century. However, if some of the additionally assumed resource of 153 million tonnes (much of which lies in the realms of surmise) could be added to the reserve total the situation could be eased. New locations are not envisaged in making up any shortfall. The situation is such that a much more concerted effort in the field of recycling is now afoot with automobiles offering a far from realised potential for scrap.

Another way of meeting supply deficiencies is, of course, through substitution. In the case of zinc, where it is used for iron and steel products, competitive materials using cadmium or more particularly aluminium electro-plating are gaining ground as are the use of ceramic or plastic coatings. Where zinc is used in the chemical industry significant levels of substitution have already been achieved whilst the use of zinc oxide in the photocopying market is diminishing rapidly in favour of plain paper copiers. The use of zinc alloys in die casting is the one area where substitution has not been achieved and which continues to consume large quantities of zinc, though successes have been achieved in utilising zinc in this context more efficiently.

With lead the battery market is the major source of demand and the use of alternative metals has only been effective in the case of certain specialised modes; certainly there has been little to choose on the basis of price. However, with the development of longer life conventional batteries the rate of consumption in this area might have fallen had it not been for the increased use of electrical vehicles in urban areas, the power source for which remains the lead battery. Reductions in the use of lead in petrol (down by 50,000 tonnes in 1980 alone) and its discontinuance as a constituent of interior paints will ease the supply situation, and as a covering for cables and as a material used in the construction industry, plastics and aluminium are cheaper and more easily handled materials. Lead piping has also been substituted for by the use of copper which is not only lighter, but does not give rise to problems of toxicity in areas of acid water supply.

Copper is mainly used by the electricity and electronics industry where its contribution has been highly significant. The greatest opportunities for its replacement have occurred in the power, wire and cable sectors where some 40 per cent of insulated cable and over 90 per cent of bare conductor applications (for example, the UK national grid system) are now provided by aluminium. Indeed countries such as France, India and the USSR forbid the use of copper where aluminium can be used. However, it has to be remembered that if the trend towards placing power cables underground on environmental grounds becomes more prevalent, then the substitution of aluminium could be halted in spite of the price advantages of the latter; (aluminium was cheaper per tonne of refined metal than copper in 1980). Aluminium can also substitute for copper in some heat exchange applications, particularly in the production of refrigeration equipment, and considerable potential exists for the more general use of aluminium instead of copper. All the same in some applications the properties of copper make it irreplaceable, particularly those situations where its dependability as a conductor of electricity or heat and its resistance to corrosion are vital factors.

Of the metals discussed in this section tin is by far the most expensive, averaging over $14,000 per tonne in 1980. Because the major uptake of the metal (over 40% by output) has been by an industry whose end product, cans, are very low cost containers, producers have sought to decrease the amount of tin needed to coat what is a thin steel can, albeit at a time when demand for their products was rising rapidly especially from the less developed countries. Having moved from coating steel with tin to the electrical plating process, typical tin plate is now 99.5 per cent steel and 0.5 per cent tin. It was the achievement of these low but very effective tin coverings that helped to ward off substitution. Coupled with the fact that recycling tin plate is not a viable economic proposition though it is technically possible, more recent attempts to substitute either plastic or aluminium containers have since 1974 been thwarted by the large energy requirements of the aluminium smelting industry and the input of petrol-chemicals to the plastics industry. As a result of steep rises in oil and fuel costs of all kinds, neither aluminium or plastics can now be seen as having a major substitution impact. This is not to say that other forms of packaging likely to be more favoured in cost terms are not being experimented with. In the USA cans coated with chrome oxide have been tried. However, the achievement of the two-piece light-weight steel can that uses less steel but is still coated with tin seems to be the product gaining ground in terms of costs effectiveness. A market that had gone for the oxide coated can for a time is rapidly reverting to tin plate.

The use of tin in solder alloys, which represents about 25 per cent of the tin market, appears to be untouched by threats of substitution. In the growing printed circuit market, no satisfactory alternative process to soldering has been found. Other uses, including the making of bronze and tin–copper alloys are maintaining their levels of usage. In striking contrast to zinc, where it has been largely phased out of use in the chemical industry, specific attempts have been made to get a larger foothold in that market for tin. Organotin compounds are now used in the construction industry for treating timber, as a stabiliser for PVC, and as a non-toxic substitute in agricultural sprays replacing compounds containing mercury. This expanding market can be accounted for by the fact that the percentage of tin demanded in these chemical usages is no greater than that in tin plate. However in most of these new roles tin has substitutes and clearly the metal could become most vulnerable should profit margins and/or production costs rise to unacceptably high levels or should other considerations, possibly political, raise the comparative price of tin.

The minerals discussed in this section are of basic importance to the world economy. Moreover, there is an essential separation, especially in the case of tin and copper, between the less developed countries as producers (which may indeed rely heavily on the exports of these minerals for the inflow of much needed foreign exchange) and the developed countries as consumers (which equally rely on the availability of these raw materials at stable prices for industrial production). It is not therefore surprising that at least some attempts have been made to even out the inevitable market fluctuations in supply and demand and therefore price.

The longest standing of such arrangements is represented by the International Tin Council (ITC). As an organisation which represents both producers and consumers it has sought through buying for, and selling from, a buffer stock of tin to keep the price of the metal between well-defined upper and lower price limits in order to satisfy reasonably the needs of both sides. Unfortunately its history is one of only partial success because of the presence outside ITC agreements of the major producer/consumers of China and, until recently, the USSR and the world's largest consumer, the USA, which joined as late as 1976. It is the USA, buying and selling tin on behalf of its large strategic stockpile (at least ten times that of the ITC buffer stocks over the period 1960–80) that has been one of the major market factors.

Coupled with the obvious fact that the ITC has been of itself quite unable to have any influence over the fate of the major world economies either in terms of short-term business cycles or longer term economic growth, its capacity to act in any meaningful sense on behalf of its client countries is limited. Against such a background it does not augur well for the establishment of a body to look after copper as occurred in 1976. In this instance, however, the object of the exercise is to protect the interests of copper producers only and it is this group of nations that have formed the Council of Copper Exporting Countries (CIPEC). Unfortunately, it does not include all the major exporters and like OPEC there was disagreement amongst its members about production limits. Where one or more countries are dependent on maximising their inflow of foreign currency through their only export of any significance the temptation is to produce as much as possible in the hope that it can keep ahead of falling product prices. The copper producers of Zambia and Zaire are cases in point.

Undeterred by this and other experiences, a United Nations study group has been looking at the creation of an organisation to bring together the producers and consumers of zinc. Even more ambitious is the idea put forward by the United Nations Committee on Trade and Development (UNCTAD) for a proposed Integrated Commodity Scheme. This would not only attempt to stabilise commodity prices for all the major minerals but also to

147

satisfy the needs of the less developed countries for technology and capital to develop their own resources for themselves, rather than rely on the influx of foreign capital and firms. The whole question of the finance and development of future mining enterprise was dealt with in greater depth in Chapter 4.

### 6.2.4 Gold; silver; platinum metals

*Gold*

Gold's distinction as a metal lies in its value as an industrially useful commodity, its virtue as a decorative metal and its importance as a means of exchange in financial markets.

Its history as a form of currency derives from its scarcity and the costs involved in its extraction thus conferring on it a high unit value. Furthermore, it is virtually indestructible (it cannot be attacked by acid or other corrosive agents), it is portable and there are no opportunities for it to be over-produced. Although no longer used as an official form of tender (though coins are still minted along with medallions as a form of investment that will maintain its value), governments along with agencies such as the International Monetary Fund hold gold in order to support their financial transactions on international markets. Their purchases or sales of the commodity along with those emanating from the USSR since it entered the world market in 1970, can affect world gold prices. However, the factor that has most influenced the large fluctuations in the value of the metal (since international fixed price agreements were abandoned) has been its wide appeal as an investment vehicle in periods of particularly high inflation.

The acquisition of gold as bars, coins, medallions or even in the form of jewellery by private individuals can then push up prices because of the supply inflexibility of an already scarce commodity. As far as jewellery itself is concerned (and over 60% of all gold is utilised for this purpose) its usefulness goes well beyond that of scarcity value. Not only has it decorative appeal and will not tarnish or corrode, but it can also be drawn out as a very fine wire and can be beaten to a fine leaf no thicker than 0.0001 mm when it still retains its malleability.

Of its industrial applications (and some 15% is used in this context in the USA), non-corrosiveness is again a significant feature, together with its readiness to alloy* with other metals. These qualities make it useful in dentistry, but as it is also a very efficient conductor it is used in electronics, particularly where high quality contacts are required and in the manufacture of bimetallic strips.

Gold deposits are found in a wide range of rock types and in the formations that range in age from pre-Cambrian to late Tertiary. Many deposits

---

*To express the weight of gold in an alloy, the term 'karat' is used and means one twenty-fourth part. Thus an 18 karat alloy contains 18/24 or 75 per cent gold.

which contain the metal are located adjacent to acidic igneous intrusions and are to be found as lodes. However, weathering and erosion cause gold in the metallic form to be released from such primary deposits and to accumulate as nuggets or as grains to form residual or stream placers. Thus although the richest lode deposits have been located in small fissure veins with a largely quartz gangue, much greater quantities of gold have been obtained from large moderate-to-low-grade deposits.

Placer mining was originally a significant source of the mineral as in the gold rushes of the Yukon, Alaska and California in the middle of the last century. However, much of the world's more recent production has come from deep narrow veins such as those of South Africa where the problems of high temperatures, humidity and extreme rock pressures all create a high-cost mining situation. This has inevitably had the effect of ensuring the maximum efficiency of gold recovery from the ore body and a combination of cyanidation, amalgamation, flotation, gravity concentration and smelting at integrated plants has ensured the attainment of rates between 92 and 95 per cent.

South Africa dominates world gold production (Fig. 6.11). In 1980 it was responsible for producing over 22 million troy ounces,* more than 71 per cent of those supplies from the non-communist countries. At least seventeen other nations around the world produced significant amounts of the metal, frequently as a by-product of copper operations (over 50% of US gold is produced in this way). However, the next largest share, outside the USSR, came from Canada at a mere 5 per cent. This was followed by Brazil (under 4%), the USA (under 3%) and Australia (less than 2%). Soviet gold production has been estimated at well under half that of South Africa. Thus, notwithstanding the importance which the recycling of the mineral must clearly have within the market (in 1980 some 13% of demand was met in this way), the propensity of two countries both with distinctive political ideologies to impact on supply and demand relationships is evident. Equally clearly the capacity to break out of this situation in the short term appears remote when the most recent USBM survey of world reserves indicated that 51 per cent of gold reserves are in South Africa, 24 per cent in the USSR, with the USA and Australia both having some 4 per cent each.

The high levels of gold prices (rising to $850 per oz in January 1980), which were maintained throughout 1980, have been responsible for keeping gold exploration activities at an all-time record level both at known sites and in virgin areas. Although new projects coming on stream in many countries have merely offset the general trend at extant mines to work lower grades, there are potentially new large gold production operations in the North West Territories and Ontario in Canada, and there is the possibility of further significant alluvial production in the Amazon basin of Brazil and

---

*Gold, silver and platinum metals continue to be traded in troy ounces on world markets. 1 troy oz = 31.1034 grammes.

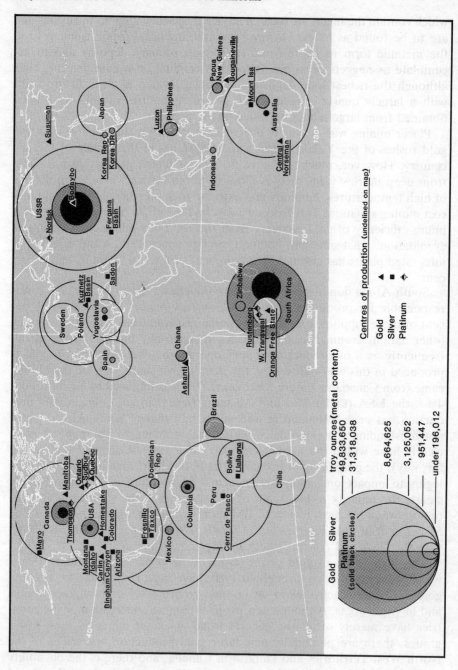

*Fig. 6.11* Gold, silver, platinum: output by country with chief centres of production, 1980. *Source: Mining Annual Review, Minerals Year Book*

new discoveries in the USA (Nevada and California), Western Australia, Tasmania, Papua New Guinea, Ghana and Chile. Perhaps the USSR offers the greatest future potential since a good deal of emphasis is being placed on metallogenic studies by the Soviet government in large areas of unexplored territory. Even though it seems unlikely that more than marginal adjustments in the location of future supplies will be achieved in the immediate future, the longer term outlook to the end of the century suggests otherwise, with a reduction in South African supplies to between one-third and a half as the trend in average ore grades continues the downward pattern of the past decade.

In terms of future demand, preventative dental programmes are likely to have some influence, whilst in the electronics industry increased efficiency in the use of gold and some substitution (tin-nickel, palladium or silver) will only have a marginal impact on a rapidly expanding industry with rising levels of demand. However, more crucial will be the desire to own gold bullion which will be related to expectations regarding inflation and other international monetary uncertainties. If world economic conditions improve, the propensity to hold gold, such an important phenomenon in recent years, will be much less. Given such uncertainties the USBM has somewhat bravely forecast an annual world growth rate of 1.4 per cent for gold up to the end of the century. Should this prove to be accurate it is possible, given current knowledge of world reserves that by the year 2000 there might be a 25 per cent shortfall in primary production which could only be made up from bullion stocks and/or recycling. This is plainly an important element in gold supply. An alternative scenario suggests that world reserves as currently known can meet 90 per cent of likely requirements. Moreover, since resources could cover twice the forecast requirements provided the gold price ran well ahead of mining costs and technological and environmental problems could be coped with in relation to the working of lower grade ores, the shortfall could be met by part of such resources becoming reserves.

*Silver*

As with gold, silver is an important industrial mineral, is used decoratively in the jewellery trade and in the production of silverware, and plays an important role as a means of international exchange, either because of its use in coins or as a means of hedging against inflation. Although for many years more extensively used as a form of currency than gold where its greater hardness and its lower but significant value made it useful as a coinage metal, it has not been used to support international currencies. Its commercial utilisation is, however, rather wider than gold since it not only performs similar functions in the electronic and electrical equipment fields, but it is an essential ingredient of photographic film which is by far the largest single consumer of the metal. In the USA photographic film took some 28 per cent of total consumption in 1980. It is also valued as a brazing alloy and as a solder and is a main constituent of batteries for the powering of electrical equipment where a long shelf life, temperature stability and high surge voltage under load are important. As a metal used in the jewellery trade and for the production of silverware its qualities are evident enough, though, unlike gold, it does tarnish.

Silver is found in the veins of igneous rocks as a finely disseminated metal of hydrothermal origins and, much more than gold, is won in conjunction with other minerals. Only about one-third of output is derived from mineral ore in which silver predominates, the chief source being those containing much larger concentrations of copper, lead or zinc. The method deployed to win the ore, whatever its association, varies from one ore body to another, but it is most frequently by open-pit methods and surface shafts and drifts. The recovery of the metal is by a flotation process which has the effect of removing the silver from the product of lead, zinc or copper smelting. If the mixture is one of molten silver-lead, zinc is added. On cooling the silver and zinc separate from the lead and rise to the surface where they can be skimmed as individual layers. In predominantly copper ores, the silver is removed during electrolytic refining, a method commonly deployed in recycling silver which contains quantities of other metals. Using an electrolyte of silver nitrate and nitric acid, fine crystals are produced of 99.9 per cent purity which can be remelted into bullion bars or ingots. As with gold, shipments of silver normally take this form. These go largely to producers of semi-manufactured wrought products such as rolled and extruded bars, rods, wire, sheet, foil and so on, and these in turn go to the manufacturers of the retail products.

Of the major developed countries that produce silver, such as Canada and the USA (Fig. 6.11), the tendency is for a small number of larger mining companies to undertake the smelting and refining of all domestically won ore at specific locations at extractive sources. Where the latter country is concerned and imports are also large (some 78% of requirements), such plant is also found at eastern seaboard locations to receive incoming ore. But as with other metals, the tendency is for increasing amounts of ore to be smelted and refined in producer countries even if they are of less developed status; here Mexico and Peru are no exception. Nonetheless other developed countries such as Japan, the German Federal Republic, Belgium and France, which constitute the main consumers, still maintain plant capable of processing imported ore to meet their needs for refined bullion silver. They also have secondary processors to handle recycled silver whose locations are relatively footloose.

World silver reserves, unlike gold, are not highly concentrated spatially though the American continents are favoured with the USA and Canada each having around 20 per cent, Mexico 13 per cent, and Peru 8 per cent. The USSR has a further 20 per cent and Australia 13 per cent. Of these Mexico has the distinction of being the leading producer with 1.6 million kg in 1980 or 19 per cent of world output, followed by Canada 16 per cent, Peru

14 per cent, the USSR 13 per cent, the USA 12 per cent and Australia 8 per cent.

In contrast to gold, COMECON countries have not attempted to sell silver on world markets and in 1980, as in previous years, there was a growing gap between mine production and the higher levels of consumption. It is a situation which is not helped by the dependence of silver production on the output of other metals. This shortfall has been made good not merely by the recycling of silver no longer required, but by the drawing down of world silver stocks in their various forms. Thus in 1980, to meet a demand for silver which was 30 per cent greater than production from mines, quantities of silver had to be disposed of by the US Treasury, coinage was melted down and some of the very large amounts of silver traditionally held by the people of India were made available. Although the export of such silver treasures from India is illegal, the attraction of high prices has overcome any resistance to meeting market needs.

In spite of a brief but dramatic drop in the price of silver from the record $50 per troy oz of January 1980, prices in the succeeding two years recovered and were at a level well in advance of those necessary to expand mine production and seek new workable sources of the metal. Consequently new workings are likely to come on stream in Mexico, in the USA (Nevada and Colorado) and in Sweden which currently contributes with Poland largely to Western Europe's 5 per cent of world output. Only the world recession has helped to relieve the otherwise greater gap between resource exploitation and demand with a fall in demand for such end uses as jewellery, silverware and coinage running briefly in 1980 at levels pertaining at the end of the previous decade. A dampening of general demand has affected the use of silver in electronic and electrical products, in brazing alloys and solders and in the photographic industry though in the case of the last mentioned it has only amounted to a 15 per cent fall.

Nonetheless, the USBM is forecasting an average annual world growth rate in demand to the end of the century of 2.3 per cent (1.9% for primary silver). This is based on an assessment of likely gross national product trends in the developed countries and in the less developed countries where rapid increases in the need for electrical and electronic equipment as well as refrigeration appliances are most likely to occur. However, substitution effects may be felt in these sectors and others. An example is the switch from silver halide film to the recording of optical images on video-tape and plastic discs and the discontinuance of coinage containing silver in favour of cupro-nickel.

Substantial reassessments of the silver resource base are complicated by the fact that so much of the metal is coming from lead/zinc/copper operations. But unless this is done, perhaps by finding new techniques of leaching and chemically processing low grade disseminated ores in which silver predominates, then it seems likely that demand will run ahead of primary silver availability. This means the shortfall will continue to require the reclamation of jewellery, coins, silverware plus the recycling of outworn industrial hardware containing silver and accounts for the USBM forecast that the availability of secondary sources of silver will need to expand at over 3 per cent per annum up to the year 2000. By the 1990s ocean floor deposits which are thought to be considerable could be exploited from areas such as the Red Sea, once new mining and metallurgical processes have been developed.

### Platinum metals

The last of the precious metals consists of closely related minerals referred to as the platinum group. They are platinum, palladium, rhodium, indium, rutherium and osmium and are amongst the scarcest of the metallic elements. Their cost is therefore correspondingly high although they present no particular difficulties regarding their extraction. Unlike gold and silver this group plays no part in the international monetary system, but as with the other precious metals, they have a considerable importance to the jewellery trade where scarcity, value, strength and workability make them attractive. More than this, they play an almost indispensable role in modern industry, a fact deriving from their extraordinary catalytic properties, their chemical inertness over wide temperature ranges and their high melting points. Thus in the chemical industry, these metals are used as catalysts in the making of a wide range of chemicals and pharmaceuticals. In petrol refining catalysts of platinum, platinum-indium and platinum-rhenium are vital elements in the process by which the octane rating of petroleum fuels is upgraded. Moreover platinum-palladium catalysts are used in commercial vehicles to oxydise exhaust gases (carbon monoxide and unburned hydrocarbons) thus lowering their pollution potential.

Platinum-rhodium alloys are also a feature in the hardware connected with the production of glass fibres and the group as a whole is widely used in dentistry (in orthodontic and prosthodontic devices) and in the production of certain medical instruments such as hypodermic needles and cardiac pacemakers. Perhaps even more significant is the wide variety of electrical and electronic devices which use this group of metals to take advantage of their chemical inertness and thermal stability.

The platinum group of metals is associated with basic and ultra-basic rocks. Almost all the current supplies are derived from lode deposits in these two rock types. In the basic rock the metals are found as disseminated minerals of one or more elements in the group. The metals are also often associated with nickel and copper sulphides, this being the prime reason for their working at Sudbury in Ontario and Thompson in Manitoba. In ultrabasic rocks, the platinum metals appear as discrete minerals, but where these have disintegrated and the minerals are concentrated as placer deposits, palladium is often absent. Placer deposits, however, only account for about 3 per cent of total world production and their extraction, usually by bucket-line dredging, is limited to the USSR and South Africa and the very minor producers in Alaska and Colombia. The extraction of the chief deposits of platinum minerals takes place, for the most part, underground by conventional but diverse techniques which take account of local conditions.

If the platinum minerals are subordinate to the extraction of nickel and copper (as at Sudbury), they are recovered during processing from the anode slimes that collect in the electrolyte cells of the copper and nickel refineries. If, as is the case in the major South African workings of the western Transvaal, nickel and copper are subordinate to the platinum metals, preliminary crushing and screening of the ore gives rise to some high grade concentrate which goes directly to the platinum refinery. The rest is further concentrated by froth flotation, smelted in electric furnaces to make a nickel-copper-iron matte, and blown in convertors to make a high grade nickel-copper matte containing well over 1 kg (31,103 troy ounces) of platinum metals to the tonne of matte. The electrolytic refining of the matte at a nickel-copper refinery gives rise to anode slimes of 25–75 per cent platinum group metals. These can then be removed by chemical methods achieving recovery rates of around 98 per cent for platinum and palladium though slightly less where the other four metals are concerned. Overall about 80 per cent of the platinum group of metals is recovered from ores in the Sudbury area and 75 per cent in the case of South African ores.

The above references to types of deposit and processing techniques have already hinted at some of the major sources of platinum minerals (Fig. 6.11). Their spatial concentration is even more pronounced than that of gold with once again, South Africa as the dominant supplier at 75 per cent of world output in 1980. Alternative sources are even more restricted with the USSR providing 11 per cent of total world supplies although its output is considerably greater. The only other country with a significant output is Canada which in 1980 produced 4 per cent of world supplies. With only three major companies engaged in the production of platinum minerals in the world, apart from the state organisation of the USSR, refining patterns are quite striking. Where the chief product consists of platinum minerals alone, once preliminary on site processing has occurred, the concentrates are despatched by the mining company to their own local refineries (in the case of South Africa, Johannesburg).

For one South African operator, however, which alone produces three quarters of world output, a residue of the pre-1970 situation is still apparent as some is shipped to London. Where the complications arise from extracting platinum metals from electrolyte sludges left over after the removal of copper-nickel (as in the case of the INCO workings at Sudbury, Ontario), these are sent to London for final processing. Those produced from the Norilsk working in Siberia are sent 900 km south to Krasnovorsk.

The major industrially developed countries are the key consumers of refined platinum ores. The largest importer, the USA, utilises about one-third of world supplies available on the free market and receives 60 per cent of its needs equally from South Africa and the USSR. It obtains a further 21 per cent from the UK, which because it has refining interests, supplies this

155

as refined metal. Concentrates comprise the US imports from South Africa, a form of platinum which it also obtains from Canada. Japan imports refined metal from the USSR, which, with the UK, also serves the needs of EEC countries.

In the USA about 41 per cent of consumption is used in the control of car exhausts, 16 per cent is used as a catalyst by the chemical industry and 15 per cent in the manufacture of electrical and electronic equipment. Although data on consumption is not available for other developed countries, it may be reasonably assumed that their patterns are somewhat similar with the notable exception of Japan. There 75 per cent of platinum goes into jewellery, whereas in the USA and Western European countries this figure is likely to be in the range of 2–5 per cent.

Where future demand is concerned, in calculating the likely levels of consumption up to the end of the century, the USBM has carried out its usual analysis of socio-economic indicators in all countries and considered the factors likely to create significant changes in key areas of consumption, specifically its use in petroleum refining, as an automotive emissions control catalyst, in the manufacture of glass fibre and jewellery. All of these offer some potential for alteration or substitution. In connection with the last mentioned, a well-backed campaign to increase its use in Western countries to levels currently enjoyed in Japan could well meet with some success. The figure arrived at by the USBM suggests a world average increase in consumption of around 2.6 per cent. Against a background of world reserves estimated at 1,180 million troy ozs and the use of recycled platinum running at around 10–11 per cent of total consumption, world supplies are more than ample at twelve times the forecast demand between 1980 and 2000. However 95 per cent of primary supplies of the platinum group of minerals come from two countries, South Africa and the USSR, which also hold a very large percentage of total reserves and much of that of the entire resource base though the USA and Zimbabwe may also figure here. Both South Africa and the USSR are also major producers of one of the chief metals of the group, with the former producing two-thirds of palladium. This must leave the major developed countries vulnerable to the actions of other governments whose political attitudes they may not entirely view with favour. It is then perhaps not surprising that Canada is developing new palladium–platinum capacity in Ontario whilst in the USA, Montana is attempting to utilise some small part of its sizeable but hitherto undeveloped, poorly defined and mostly sub-economic platinum resource believed to be of the order of 300 million troy ounces.

### 6.2.5 Antimony; Rare earths; Lithium

*Antimony*
The mining of antimony by open-pit or underground methods is largely associated with the extraction of other metals including lead, silver, gold or

tungsten since it occurs as a constituent, in one of its 112 forms, of these other ores. At its most prolific it never does more than share in the total value of the output of such mines. In the most complex deposits, the antimony content is so small that it is only detected at the later stages of processing.

Though frequently alloyed with lead and other metals, antimony may be considered as primarily of significance in the chemical industry not only because in this form it finds its way into batteries whose function depends essentially on a chemical reaction, but also because it has recently become important as a flame retardant in the production of vehicle body trim. Other but less significant uses of the material in its oxide form are in the making of glass, in ceramic glazes and in the manufacture of rubber where it acts as a vulcanising agent. Its close association with the production of vehicles has, however, been a disadvantage in the current world economic recession since this industry tends to be most immediately responsive to down-turns in demand for consumer goods. Moreover the battery side of the utilisation of this mineral seems likely to be increasingly adversely affected by a transference of demand towards calcium-lead units which are largely maintenance free. Thus overall world production of the most common ore, stibnite (antimony sulphide), is not likely to show any marked increase in the immediate future. Of the 61,000 tonnes produced in 1980, Bolivia and South Africa and to a much lesser extent, Australia, Canada, Mexico, Thailand, Turkey and Yugoslavia supplied nearly 70 per cent of world output with China and the USSR between them accounting more or less equally for a further 28 per cent (Fig. 6.12).

Changes in the major uses to which antimony has been put have been reflected in changing world trade patterns. Both Bolivia and China have tended to remain suppliers of concentrates which originally characterised the majority metallic outlet, though there is an increasing propensity to add value to the extracted material via smelting activity. On the other hand production from South African sources has tended to serve the oxide side of the market characterised by the growth in demand for fire-resistant vehicle trim materials at least until the onset of the economic recessions of the 1970s and early 1980s.

Both product substitution (in the case of batteries) and the fall in demand due to economic recession have prompted producers to contemplate the formation of a producers' association to regulate world prices. The success of this is far from assured since two of the four key producers, China and the USSR, not only have centrally planned economies which would inevitably preclude their participation but they are politically antagonistic towards each other.

World reserves of antimony stand somewhere in the region of 4.8 million tonnes, a figure which represents 84 per cent of the likely resource base. With forecasts by the USBM of an annual growth rate of 2.3 per cent to the end of the century, it would seem that world shortages are unlikely. Whilst

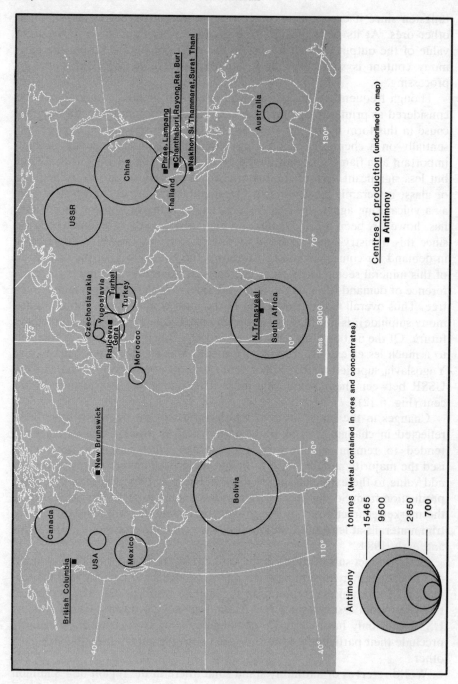

*Fig. 6.12* Antimony: output by country with chief centres of production, 1980.
*Source: Mining Annual Review, Minerals Year Book*

half of the total estimated reserves are in China, the rest are evenly distributed amongst the USSR, Bolivia, South Africa and Mexico (all in the range 8–5%) and Australia, Italy, Malaysia, the USA, Turkey, Yugoslavia and Thailand (in the range 3–2%).

*Rare earths*

The term 'rare earths' is a misnomer since it is used to cover a group of minerals that have nothing to do with earth as such. This is just a term used by early chemists to cover a range of oxides – and they are far from rare. Supplies of the minerals that largely constitute the rare earths (bastnaesite, monazite, xenotime and yttrium) are relatively widely available with most of the continents having adequate reserves, though within these a more marked concentration is observable. India has 91 per cent of Asian reserves (46% of the world total), the USA 76 per cent of those of North America and the USSR 89 per cent of European reserves. Australia and Brazil have all of the currently exploitable rare earths in their respective continents.

Rare earths have been extensively used in the polishing of glass where a high quality finish is required (for example, lenses, mirrors, television face plates and so on). But as this market has tended to decline with the development of new glass-making technologies (such as the Pilkington float-glass process), their use in the *manufacture* of glass has itself become of increasing importance. Rare earth additives prevent discolouration of the glass, minimise changes in opacity where it is subjected to radiation (as in television screens) and improve the refractive qualities of high-grade lenses. Since 1970 a demand for rare earths also has come from the iron and steel industry. With the greater production of large diameter steel tubing, it has been found that the addition of rare earth silicide can improve the rolling properties of steel sheet. Similarly the addition of rare earths in the production of ductile cast iron has improved its strength and hardness and thereby opened up new possibilities for the production of high-quality automotive and agricultural castings and ductile iron pipe for water transmission systems. In other recent applications rare earths have proved of value in enhancing the thermal tolerance of ceramics, as an addition to iron–nickel–chrome alloys which have to work at elevated temperatures, in the manufacture of high quality electro-magnets and in a number of nuclear applications. But their greatest significance in terms of current outlets is in the production of catalysts for the refining of petroleum.

The percentage of rare earths used by each of these key sectors of consumption is not of the same order throughout the developed world. In the USA the main end use is as a petroleum catalyst (41%), followed by iron and steel (35%), ceramics and glass (19%) and other uses (including magnets and alloys) 5%. In contrast, Japan's utilisation is very different because of its highly developed electronics and photographic industry. Rare earths are in great demand there for television tubes, lenses and more particularly, magnets used in electronic watches, stereo turntables, computer terminals, line printers and so on.

Rare earths (or lanthides) cover a group of some fifteen chemically similar elements of which yttrium is not strictly one. However it is usually classed with these elements because it occurs with them in nature and has similar properties. Both this, xenotime, (which is in fact a yttrium sulphate) and one of the other chief sources of the rare earths, monazite, occur together in many geological environments, but are only worked in alluvial beach sand deposits where they have (with other heavy minerals) been concentrated by the action of wind and water. Even then it is often necessary that they be exploited as a by-product of ilmenite, rutile, zircon or gold. Such deposits occur in large quantities along the coasts of India, Brazil and Australia, although they are similarly found in Malagasy, Sri Lanka, Korea, Uruguay and Argentina. Bastnaesite, the other principal source of rare earth, is found both in veins and disseminated in a complex of carbonate-silicate rock areas such as California, Burundi and New Mexico.

Most placer deposits are worked by dredging. Other disseminated and vein deposits are extracted respectively by open pit methods and underground techniques. Processing to recover the valuable rare earth minerals and to produce a concentrate can involve the use of gravity methods though electromagnetic and electrostatic techniques have been used on monazite and flotation methods on monazite and bastnaesite. Further processing may be carried out involving leaching with acid and/or roasting to produce pure rare earth oxides. These may then subsequently be reduced to metallic form if the market for the product so demands.

Most rare earths enter the world markets as concentrates from the key exporting producers (Australia, Brazil, India, Malaysia and the USSR). Of these only the USSR may be considered a consumer of consequence whilst others consist chiefly of the UK, France, the USA (itself a producer) and Japan. The last mentioned country, because of the high levels of demand from its expanding electronics industry, is attempting to offer technical help to the Chinese in the opening up of what now appears to be the considerable reserves of Inner Mongolia. Figures relating to the early 1980s indicate that in spite of a world recession, the diminished needs of the lens industry for the use of rare earths as a polishing agent and the possibility of some limited if not entirely satisfactory substitution, increased demands in COMECON and the less developed countries for television equipment and by the developed world for computers and other electronic equipment, will ensure a buoyant and growing market for rare earths. The USBM has proposed a yearly world increase in consumption averaging 6.25 per cent on the 36,700 tonnes produced in 1980 up to the end of the century.

*Lithium*
In 1980 world lithium consumption was over 24.5 million kgs, excluding the USSR, which may have utilised between 3.6 and 5.45 million kgs. Lithium is extracted from naturally occurring lithium rich brine or is mined as an ore.

The ore is found as crystals of spodumene, lepidolite or petalite in pegmatite rocks and is worked mainly by open pit methods. Ore extraction depends on the nature of the original material but is mainly by flotation.

It may be used in a number of chemical forms. As a carbonate, by far the most important role of the lithium salts, the growth in its use as part of the electrolyte in aluminium reduction cells has been notable especially in Western Europe and South America. In its hydroxide form demand continues to rise from the manufacturers of multi-purpose greases. As butylithium it is consumed by the synthetic rubber industry. In its carbonate form and as an ore it is used by the ceramics and glass industry though in both instances demand has shown signs of falling. Although lithium has a wide range of other uses, the remaining key area of consumption which involves the metal itself, binary alloys and lithium salts is that of batteries, and there the market may be most buoyant.

Lithium is not widely produced as an ore, or from brine. Only the USA, Brazil, Namibia (South West Africa), Zimbabwe and Chile are involved. However, the organisation of the secondary processing of lithium as a chemical for use by other industries shows interesting development. Where a developed country which has reserves is concerned, point sources of the production of lithium chemicals and metal tend to be wholly integrated with mining operations. In the USA, the centres of Bessemer City and King's Mountain, both in North Carolina, are not only mine locations but also have plant capable of producing carbonate. Total capacity for the last mentioned at both plants amounts to 21 million kgs per year.

Similar integrated facilities aimed primarily at the production of over 7 million kgs of lithium carbonate per year exist at Silver Peak, Nevada, but there the plant is fed with lithium brine. Such approaches aimed at increasing the value of mining production are appearing in less developed countries. Zimbabwe, the world's leading producer of low iron lithium, ideal for use in specialist glass and ceramic manufacture, no longer markets lump ore but instead offers ground products to several different user specifications, and in 1984 the Chilean government planned to open a plant producing lithium carbonate from the brines of Salar de Atacama. As a residue of the older patterns of production where the basic raw materials were exported to major developed countries for secondary processing, appropriate examples are found in Europe and Japan. The German Federal Republic continues to be the largest EEC manufacturer of a variety of lithium chemicals and catalysts, whilst in 1980 the UK brought on stream a large alkyl lithium catalyst plant aimed at supplying synthetic rubber producers in the same trading bloc. Japan, the world's largest single producer of lithium chemicals from imports of the raw material, not only supplies home markets but is a key exporter to other Asian countries.

Reserves of lithium and the capacity of plant to process ore or brine appear to be more than adequate to meet any reasonable forecasts of increased levels of demand for the future. Reserves seem to be in the region of 2.4

million tonnes of contained lithium and a further additional 6 million tonnes of resource appears to exist. Even if the high figure for the annual percentage increase in demand up to the end of the century of 5.9 postulated by the USBM is realised, total cumulative demand from 1980 may not be much more than 300,000 tonnes. China may increase its exports, but new ore production capacity will probably be developed in Chile, Canada and Zimbabwe as well as the USSR and new brine deposits exploited in Bolivia similar to those currently extant in Chile.

Future demand may well depend on the production of aluminium, lubricating grease, ceramics and rubber, yet the key factor will be a continued acceleration in the need for lithium batteries though only small amounts are consumed per unit. As these power sources have already found wide commercial and military application and can clearly outperform conventional types of battery, there is reason for optimistic growth projections.

### 6.2.6  Mercury; cadmium; indium, rhenium, selenium, tellurium

*Mercury*

These minerals are characterised not only by their high value, but also by their common primary usage in the electronics industry. Thus 40 per cent of the output of mercury finds its way into the production of batteries, electrical apparatus, lamps, rectifiers and switches. A further 25 per cent is used in electrolytic cells for the production of caustic soda and chlorine. There are also other major outlets including paints (14%), the manufacture of certain measuring instruments such as thermometers and barometers (13%) and dental supplies (5%).

In its refined state mercury is the only metal that is liquid at ordinary temperatures. It is mainly obtained from a red sulphide mineral, cinnabar, even though it is known to exist in a number of others. This is generally found in association with limestone, calcarious shales, sandstone, serpentine, chert, andesite, basalt and rhyolite. Frequently located at relatively shallow depths in areas of late Tertiary orogeny and volcanic activity, the ore can be classified into two main types according to its occurrence. It is either found as a disseminated ore in which the cinnabar has impregnated a fine grained or highly brecciated gangue, or in fissures and cracks of the appropriate rocks.

The ore is frequently extracted using open-pit techniques but most often by underground methods. Primary processing involves ore crushing and sometimes screening. The mercury is then extracted from the ore by heating in retorts or furnaces. This liberates the metal as a vapour which after cooling is collected as a condensed liquid metal. Ore as such is not generally exported from producer countries, the material being sold to importing countries as a liquid by the flask (76 lbs or 34.5 kgs).

The main producer of mercury is the USSR with an estimated output of 60,000 flasks, representing 32 per cent of the world total in 1980 of 190,000 flasks (Fig. 6.13). Of this production figure, the USSR, itself a major con-

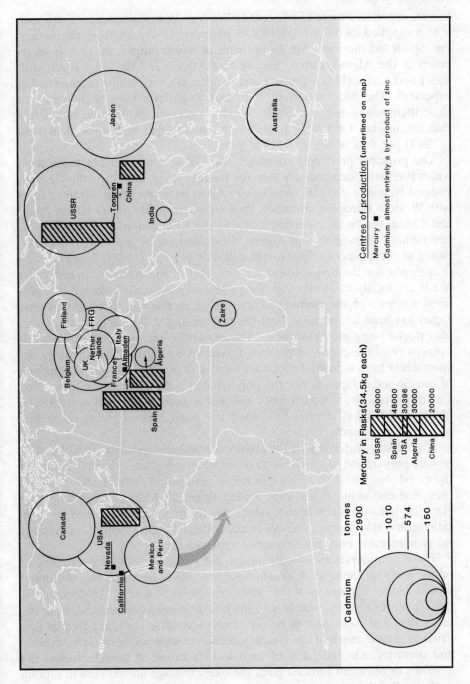

*Fig. 6.13* Mercury, cadmium: output by country with chief centres of production, 1980. *Source: Mining Annual Review, Minerals Year Book*

sumer, exported a mere 15 per cent to Western European countries although it also supplied its other COMECON partners. Of the other major producers, Spain led the way with 25 per cent of world output, all of it from its mines in the Almaden area, followed by Algeria and the USA (each with 16%) and China (11%). Canada and Mexico are very minor producers and exporters. The once important Italy and Yugoslavia have been forced to close their mines due to excessive production costs and in spite of the fact that the market for this high priced commodity had held up well in 1980 (c. $400 per flask at the end of the year).

The principal producing countries, the USSR apart, are not those from which the major demands for mercury come; that is, the USA, the German Federal Republic, the UK and France. Apart from the USSR's trading links with Western European countries, their suppliers otherwise are mainly Spain and China, the latter exporting solely to the EEC countries. The USA imports mainly from Spain and Algeria. One other major importer, India, continues to consume considerable quantities of mercury (10,000 flasks in 1980) though not for the major world market, that of the electrical industry. Its needs are totally bound up with the production of caustic soda but it is precisely because of the banning of the use of mercury in that industry that Japan has been temporarily forced out of trading in this metal with its surplus stocks being used elsewhere. The reason for this ban derives from the propensity of mercury to have an adverse impact on the air and water environments from industrial processes such as those involved with the making of caustic soda. Japan is not the only country with this concern. The USA Environmental Protection Agency now limits mercuric discharges from such plants whilst it has effectively banned the production of paints containing mercuric compounds.

A generally wider regard for pollution problems resulting from the electrolytic preparation of caustic soda could adversely affect overall demand for the metal, but whilst there are alternative production methods that could be used, the end result is not such a pure chemical. Moreover, as far as the electrical industry is concerned, substitutes for mercury are few and far from satisfactory, thus leaving such areas as paints and dental supplies as the only really vulnerable ones for alternative replacements.

These considerations of substitution are pertinent to any assessment of future levels of demand although there are others including the likelihood of still further expansion of the electronics industry and of the demand for battery power sources especially in the less developed countries between the years 1980 and 2000. Also to be taken into account is the real potential that exists to recycle much of the mercury used in electronics, instrument making and dentistry. On the basis of such a wide range of considerations, the USBM's most recent forecast gives the likely average annual growth rate up to the end of the century as 2.3 per cent. With total world reserves currently lying at around 4.5 million flasks, distribution of deposits by size very much directly related to current production levels and no importing nations too dependent on a single supplier, problems of supply and demand seem unlikely to arise for the immediate future.

## Cadmium

Cadmium shares with mercury two significant attributes. First, it performs an important role in the electrical and electronics industry. As an electroplate it not only affords the host metal protection against corrosion, but it offers ease of solderability and low electrical contact resistance. As a component part of batteries (along with nickel) its recently recognised qualities of permitting high rates of discharge yet imparting long life have led to a world increase of spectacular proportions in its use, rising from 6 per cent in 1960 to over 20 per cent of total output in 1980. In Japan batteries now account for 40 per cent of cadmium consumption. In all, utilisation of cadmium in the electrical and electronics sector amounts to 60 per cent. Second, cadmium like mercury, if not handled with great care during its processing and with circumspection in its end use, has a potential to impact adversely on people and their natural environment. It is this latter problem (discussed in Ch. 8) that may ultimately influence levels of utilisation and which may cause a diminution in output long after markets have recovered from the recession of the 1970s and early 1980s. However, it should be remembered that the propensity to produce cadmium is closely related to levels of zinc production.

Although cadmium ores exist in their own right, they are not commercially exploited. About 95 per cent of primary cadmium is won as a by-product of zinc, a metal with which it is usually associated in the ore. The remainder of primary cadmium production arises from lead and copper smelting operations. In addition a small amount of secondary cadmium is produced from dusts, slags and other residues resulting from the production of the primary metal and from wastes, or scrap products from other sources.

Because of its associations with zinc, deposits of the metal are widely dispersed. Production predominates in the non-Communist industrially developed countries (Fig. 6.13). Twelve producers are significant, including the six chief consumers. Japan (with 16% of world output in 1980), the USA (14%), Canada and the German Federal Republic (9% each), Australia (7%) and France (6%). Of the five less developed countries, none may be considered of great significance, the most outstanding contributing a mere 6 per cent of the non-Communist world's output (Mexico and Peru). Only India ranks as a consumer utilising its total output in 1980. The only other cadmium producer of significance but not so far discussed is the USSR. Although its exact output is not known and does not appear on world markets, it is thought to be approaching that of Japan whose 2,209 tonnes out of a free market total of 14,200 tonnes was the single largest.

Recovery of cadmium during the processing of zinc bearing ores follows

165

a variety of procedures. Fume dust collected during the roasting and sintering of zinc concentrates is upgraded from its 10 per cent cadmium content in a kiln or reverberating furnace. In the electrolytic method of producing zinc, the roasted concentrate and the collected fume dust are taken into solution with sulphuric acid and a zinc cadmium sludge precipitated by adding zinc dust. The final recovery stage involves the dissolution of the cadmium bearing feed material followed by various purification and cadmium displacement steps that may involve electrolytic or electromotive treatment. Because the winning of cadmium ores is mainly in the hands of developed countries which utilise their own output, the processing and refining of the metal just described takes place internally. The export of cadmium from less developed nations has been as part of their external sales of zinc concentrates. However, moves are afoot in major producers such as Mexico and Peru to process zinc ores at home thus exporting the metallic products of zinc and cadmium.

According to the USBM it seems likely that demand for cadmium could show an annual growth rate of around 2 per cent on a world basis to the end of the century. The wide spread nature of known reserves indicates that even if there were spatial changes in the production patterns of cadmium and its related mineral zinc, it is unlikely that any shortages could arise for consumers wherever they were located and whatever the prevailing political situation.

Of the inputs to the resolution of a likely future level of demand for cadmium year on year, undoubtedly the most questionable is that of the likely response of nations to the pollution problems created by processing and industrial utilisation of the mineral. A strong possibility exists that problems pertaining to its use in some market sectors will create a down-turn in its consumption especially in those areas where it can be most easily substituted for (that is as a pigment, or a stabiliser in plastics). Uncertainty lies in the extent to which this will be compensated for by the growth in that market where these problems may be adequately contained. As for its use in batteries which is a major growth sector, it is possible to design manufacturing processes that avoid the industrial hygiene and pollution difficulties that cadmium can create.

## Indium, rhenium, selenium, tellurium

Apart from mercury and cadmium these are four other minerals associated with the electronics and electrical industries. They are produced only in small quantities and as a by-product of the winning of a more important metal. Indium occurs in lead–zinc ores and also in some tin, tungsten, manganese, lead and copper ores. Rhenium is obtained from molybdenite in porphyry copper deposits, whilst selenium and tellurium occur as trace elements in certain sulphide ores of copper and lead. By-product processing in each case resembles the methods deployed for cadmium.

Indium imparts several properties to solders including better wettability and reduced leaching of certain precious metal substrates. The use of this metal in the production of fusible alloys in the optical industry is its most important single function. Selenium's semi-conducting properties make it of use in selenium rectifiers as well as in the increasing market for low-maintenance car batteries. Much of the demand, however, has been attached to its significance as a photo-receptive medium utilised in the field of photo-copying. It is also used in alloys.

Tellurium is mentioned here, not so much for its present levels of util-isation largely in the metallurgical industry, but for its potential value in the conversion of sunlight to heat and application in domestic solar heating systems.

Rhenium is useful in the production of heating elements, X-ray tubes and electrical contacts, though again its major use is elsewhere as a catalyst in the production of lead-free high-octane fuels.

Of the four minerals, rhenium is produced in by far the smallest quantities with a total in 1980 of only some 8000 kg. Of this 29 per cent of world output is obtained from Chilean molybdenum ores, the greatest part of which is shipped to the German Federal Republic where rhenium extraction takes place. A further 23 per cent comes from Canadian sources and 17 per cent from the USSR. In both countries locally won ores are processed internally. Of the remainder, Zaire exports molybdenum concentrates to Belgium and Luxembourg for the extraction of this mineral, whilst rhenium is processed from imported concentrates in Sweden and the UK. The USA is the only other country producing this metal from indigenous ore but production fig-ures are not available.

Production statistics for the USA are also unavailable for indium though it is known that much of its output is obtained from ores imported from Bolivia and that production levels represent a large percentage of the world total. Other leading producers are Japan and the USSR, each with 6.22 mil-lion grams per year from the rest of the centrally planned economies, other than the USSR. Canada, once a leading producer, has a much reduced output due to a sharp decline in the indium content of its lead–zinc ores. World trade mainly consists of exports from Canada, Peru, the USSR and Belgium, and imports by the USA taking 40% of world output, and the UK and Japan the rest. This reflects the needs of industrial countries which are without indige-nous supplies, or whose needs greatly exceed their own capacity to refine the metal. The USBM estimates world production of selenium in 1980 to be 1,450 tonnes, primarily produced by refineries in North America, Japan and West-ern Europe, whilst the output of refined tellurium from the market economies is 245 tonnes with around half originating from the USA.

Future supplies of these metallic ores are largely determined by the avail-ability of those minerals of which they are a by-product, though in some rare instances the expected percentages of the secondary mineral can decline un-expectedly. In terms of demand, indium seems likely to move ahead strongly in spite of current recession, as its major user is the electronics industry. As a catalyst for petrol refining, as well as in other uses, rhenium could be subject to substitution though in its role as a catalyst reprocessing could ultimately constitute a secondary source of supply. In the case of selenium and tellurium, their unique properties suggest that annual increases in demand may run at around 4 per cent for the immediate future. Even so for these two minerals as well as the others under discussion here, calculations concerning availability by the USBM do not indicate overall problems of shortage when considered against the likely resource potential except at a localised level. It seems unlikely that such a mismatch could not be resolved through normal international trade.

### 6.2.7  Uranium

It will be apparent in Chapter 7 on carbons and hydrocarbons that the emphasis being placed on coal as a key source of energy in place of oil provides only a short to medium-term solution to a long-term problem.

Coal must be looked upon as a bridge linking the present with new forms of energy technology. Of the new forms the only one that appears to offer any possibility in the world's production of energy is nuclear power. As with coal, its primary aim will be to provide the means by which electricity is generated. Instead of burning a fossil fuel to drive steam turbines which in turn power dynamos, however, atomic reactors which use enriched radio-active uranium as their fuel input will be the source of heat.

The main problems which beset the production of energy in this way do not derive immediately from difficulties concerning the availability of the basic fuel, uranium, though it has to be acknowledged that stocks of this resource are finite and could begin to run out early in the next century if fresh reserves are not identified. Nor do they derive from cost since studies carried out by the United States Department of the Environment show that a KWh from a nuclear plant costs about 7 per cent less than a KWh produced from a coal-fired plant, whilst specific investigations of coal-fired power installations in the UK and Ontario have indicated a mark up on generating costs of 50 per cent and 100 per cent respectively on their nuclear counterparts. The more immediate difficulties stem on the one hand from the claims by environmentalists that this means of generating power is unsafe and opens up the possibility of major radiation contamination disasters affecting both people and ecosystems, and on the other, from the many arguments that are currently taking place about the respective merits of different kinds of reactors and their further development.

On the first point, there has been a wide acknowledgement of the potential dangers following the accident at Three Mile Island in the USA in which radio-activity leaked from the plant and of the need to be certain about the efficacy of safety measures. This has been recognised in that country to the extent that ten proposed plants with a generating capacity of 10,000 MW were cancelled in the aftermath of that event. Perhaps the most serious doubts have been expressed in the German Federal Republic where the initial introduction of nuclear power has been considerably retarded by violent anti-nuclear demonstrations and in Sweden where after much agonising the future of the nuclear programme was the subject of a national referendum. Concerning the second point, different nations with developing nuclear energy programmes have approached the question of thermal reactor building in different ways using alternative solutions to the essential problem of generating energy in the form of heat. This has led to the appearance on the scene of competing systems such as the American light water reactor and the British advanced gas-cooled reactor and arguments concerning the effectiveness and comparative safety of each. But perhaps of ultimate concern will be the achievement of breeder reactors, of which prototypes are now being built. These will, as a by-product of nuclear reaction, produce more nuclear fuel thus ultimately obviating the need for the production of uranium, the material which fuels and is consumed by the present generation of working reactors. However, since uranium is clearly the key to nuclear energy production from the present range of reactors and will continue to be so probably until the end of this century, it is to this mineral that attention must be given.

Uranium is amongst the less common elements. In terms of its average distribution in the earth's crust it is present at levels of around two parts per million but to be exploited economically concentrations have to be between 0.10 and 0.30 per cent. Although it is always combined with other elements in its parent rock material, 70 per cent of the known deposits occur either in sandstones or pre-Cambrian quartz and pebble conglomerates. Only 20 per cent of the mineral exists as veins and the rest is to be found in phosphate deposits, shales and igneous rocks. It is usually extracted by open-pit methods though veins are worked using conventional underground techniques. After crushing and grinding it is commonly leached from the parent material using sulphuric acid and is precipitated from the acid solution using ammonia, magnesia or lime. The concentrate is then refined and converted to a hexafluoride product which is gasified for subsequent enrichment, a process which increases its degree of radio-activity. It is then converted to a fuel form for use in the manufacture of nuclear fuel elements.

Levels of uranium production for 1980 are shown by chief producing country in Fig. 6.14 which also makes it apparent that output is largely concentrated in the hands of the western industrialised economies with the notable exception of Japan. The amount produced rose by 16 per cent from the previous year but perhaps the most important feature of production in that year was the rapid expansion of Australian output with a whole new complex of mines coming on stream in Queensland. Australia has contracts with nuclear utilities for the supply of this mineral to Japan, the German Federal

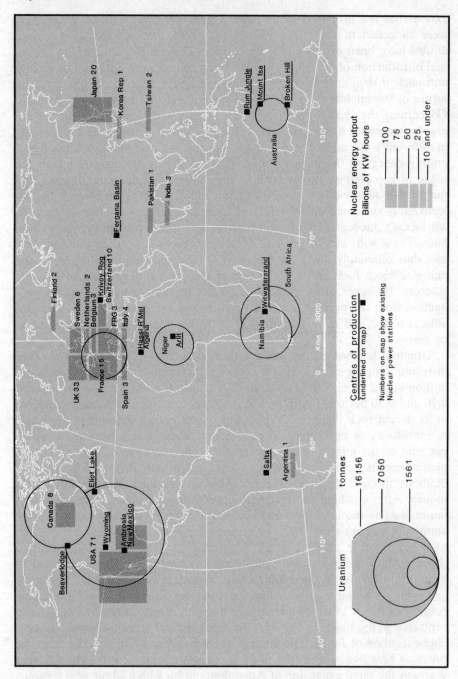

*Fig. 6.14* Uranium, nuclear energy: output by country with chief centres of production, 1980. *Source: Mining Annual Review, Minerals Year Book*

Republic, Sweden and rapidly industrialising South Korea. South Africa's importance as producer is also increasing and output rose by over one fifth in 1980 with overseas sales especially in Japan and West Germany. Of the two less developed country producers, both Niger and Namibia supply most of their output of uranium to France which itself has a modest uranium mining industry. Like France, both the USA and Canada not only produce the mineral but also consume it. Although the former has export markets in Western Europe and Japan, cut-backs in its own nuclear programme in 1980 for reasons already mentioned meant adding to its commerical stock-piles. Indeed the overall contribution of nuclear power to the total energy needs of the USA actually fell back modestly in 1980 to a figure of 12%. Canada, on the other hand not only exported to the same individual countries as the USA, plus South Korea, but also contributed to an increase in its own nuclear power generation of some 5 per cent on the previous year. Ontario is the main uranium producer and has a nuclear generation capacity which meets 30 per cent of its energy needs, but Saskatchewan is of increasing importance with several new discoveries of uranium made in 1980. In terms of the future supply of uranium which can be made available at low cost, Canadian and Australian sources may well have a distinct advantage over the other main producer, the USA.

Of the consumers of uranium, all have shown considerable increases in their demands for the mineral in the late 1970s. In 1980 alone some of those countries achieved spectacular increases in the percentage of energy generated by nuclear means over the previous year. France's increase was 53 per cent bringing its reliance on nuclear energy to 22 per cent of its total needs; Japan's increase was 30 per cent with its nuclear contribution raised to 13 per cent of overall energy supplies. Sweden and Switzerland, already leading countries in terms of their relative dependence on nuclear power (both at around 27%) expanded their programmes by 27 and 21 per cent respectively in 1980 compared with 1979. Although the more impressive levels of increase for individual countries do not necessarily reflect the paucity of other energy sources, it must be noted that the Japanese expansion is a result of the adoption by its government as official policy of a plan to reduce its dependence on imported oil from 73 per cent to 50 per cent by 1990. Such acts of deliberate policy can be fed with other data into short- and long-term prognostications about expanding nuclear power requirements. The low and high forecasts of nuclear energy requirement indicated in Fig 6.15(a) and (b) will be dependent on assumptions about the levels of economic growth and the availability of energy alternatives and their cost. Whether the requirements can be met will depend to some extent on the availability of uranium. Whilst Fig. 6.15 (b) indicates this as a finite resource, its life as a fuel for the current range of reactors can be prolonged by known techniques of reprocessing spent fuel rods which, after a time, become no longer effective in nuclear reactors even though they still retain much uranium. More

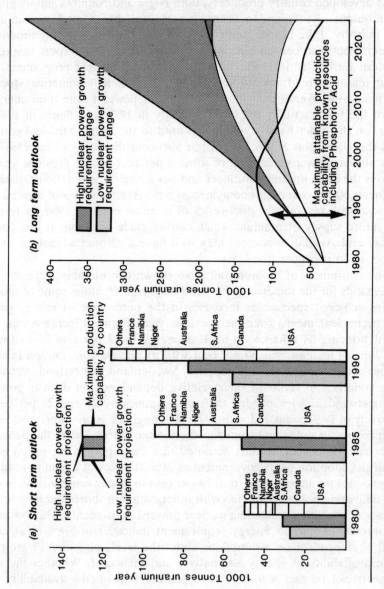

Fig. 6.15 World uranium supply and requirements. *Source: Mining Annual Review* (1981)

than this, if breeder reactors can be successfully introduced for the production of electricity (currently a more likely scenario in the time spans indicated in Fig. 6.15 (b) than the achievement of viable nuclear fusion reactors) before the demise of workable uranium sources then the potential for nuclear energy is largely unlimited. However, the boldness of such a statement must be constrained by considerations of safety and the evaluation of the environmental 'risk' factor as atomic energy plants proliferate.

# CHAPTER 7

## Key industrial minerals – carbon and hydrocarbon fuels

### 7.1 Oil

Throughout the 1960s any examination of the energy utilisation of a developed country would have indicated unequivocally the increasing role played by oil as a source of energy. Right across industry the change from coal to oil was in evidence. New power stations began to use heavy fuel oil as boiler fuel. Industry saw the advantages of using a fuel that was clean, easy to handle and store. The railways of Europe and North America switched from coal burning locomotives to diesel electric trains on a large scale. The movement of goods via diesel-powered road vehicles expanded rapidly. Civil aviation using kerosene developed on a scale that could not have been imagined two decades earlier. Personal affluence led to a boom in the use of the petrol engined private car and in the installation of oil-fired central heating systems.

To cite but one striking example of this change, in 1960 oil met 29 per cent of energy requirements in the UK, the rest being almost wholly supplied from coal. By 1973, immediately before the first oil crisis, oil was meeting almost half of the energy needs in a market shared by natural gas and nuclear generated electricity as well as coal. It was also in the period after 1960 that oil began to perform another invaluable industrial function. Once the basic chemistry of hydrocarbons had been applied for the first time in a commercial context, it rapidly replaced coal as a basic feed stock for the production of many valuable chemical compounds.

Thus it was that by the early 1970s industrial activity in the developed countries had become inextricably linked with oil. For this reason it is not difficult to understand why the price rises effected by the major oil cartel, OPEC, in 1973 and 1979, together with its ability largely to determine levels of supply, were primarily responsible for pushing the world economy into two fairly severe recessions. Since 1973 the response of the major industrial nations has been to switch to other forms of energy over which they may more readily exercise some control, to reduce levels of energy consumption via conservation techniques and the more efficient use of oil, and to find and develop other sources of oil in countries which are not members of OPEC. Certainly the impact of increased OPEC oil prices has been both to stimulate exploration and to encourage the means by which deposits of oil that were once considered uneconomic might be brought into production.

But whilst such activities have had to be undertaken within those sedimentary rock basins in which oil, formed from the residues of decomposed organic matter, has been known to accumulate, the oil companies have avoided those associated with OPEC (primarily Saudi Arabia, Iraq, Kuwait, Iran, Abu Dhabi, Quatar, Dubai, Nigeria, Libya, Algeria and Venezuela). Exploration activities have ranged from ventures in the North Sea to accelerated levels of drilling activity in the United States and Canada. The emphasis has been on increasing reserves in areas of known political stability which are sympathetic to those of the industrialised countries (see Table 7.1). However, in the USA the extra incentive which resulted in record levels of drilling and the first additions to US reserves in ten years, was the promise of the decontrol of domestic oil prices and the possibility of higher profit margins. Provision also was made for the first time to make large areas of Alaska's potentially productive oil and mineral areas available for exploration and drilling. In the UK sectors of the North Sea, the exploration of hitherto untried areas continues, held back only by the possible imposition by the British government of a supplementary revenue tax on any subsequent production. Similar constraints occurred in Canada during the early 1980s. There, though exploration tests in Hibernia and West Pembina have been promising, the date for bringing any wells in these areas on stream has been delayed by uncertainty regarding future national/provincial taxation arrangements.

The significance of taxation as a constraint on potential mineral extraction activities, explored in Chapter 4, influenced the development of the Alberta tar sands, said to contain 1,300,000 million barrels of crude oil (Outtrim and Evans 1978). Both *in situ* and steam recovery methods have been perfected enabling the four main Cretacious basins (Peace River, Wabasca, Cold Lake and Athabasca) which contain vast areas of sandstone, averaging an oil saturation level of around 70 per cent, to be exploited.

Although at 1980 levels the real cost per barrel of extracting oil from land-based sedimentary anticlines, where much of the oil is driven out of the well under its own pressure, was not more than $2, that of the much more capital intensive tar sands was considerably greater. This is not least due to the problems of getting the parent sandstone to release its oil and sweeping it into reservoir areas where it may be collected and then pumped to the surface. Amongst the more conventional oil deposits, the North Sea has proved to be amongst the most expensive to work. Nevertheless the cost per barrel of oil recovered from that source is still less than the forecast for the tar sands whether the most expensive of the production areas (Cold Lake) or

Key industrial minerals – carbon and hydrocarbon fuels

*Table 7.1* Oil Reserves 1982

| | Oil (*1,000 bbl*) | | Oil (*1,000 bbl*) |
|---|---|---|---|
| *Asia-pacific* | | *Western hemisphere* | |
| Australia | 1,622,077 | Argentina | 2,590,000 |
| Brunei | 1,240,000 | Brazil | 1,750,000 |
| India | 3,416,400 | Chile | 760,000 |
| Indonesia | 9,550,000 | Colombia | 536,000 |
| Malaysia | 3,325,000 | Ecuador | 1,400,000 |
| | | Mexico | 48,300,000 |
| Total | 19,153,477 | Peru | 835,336 |
| | | Venezuela | 21,500,000 |
| Others | 82,647 | United States | 29,785,000 |
| *Western Europe* | | Canada | 7,020,000 |
| Denmark | 473,000 | | |
| Italy-Sicily | 703,000 | Total | 114,476,336 |
| Norway | 6,800,000 | | |
| United Kingdom | 13,900,000 | Others | 810,730 |
| | | *Communist areas* | |
| Total | 21,876,000 | China | 19,485,000 |
| | | USSR | 63,000,000 |
| Others | 1,047,680 | | |
| *Middle East* | | | 82,485,000 |
| Abu Dhabi | 30,510,000 | | |
| Divided (Neutral) | | *Total World* | 670,189,406 |
| Zone | 5,840,000 | | |
| Dubai | 1,440,000 | | |
| Iran | 55,308,000 | | |
| Iraq | 41,000,000 | | |
| Kuwait | 64,230,000 | | |
| Oman | 2,730,000 | | |
| Qatar | 3,425,000 | | |
| Saudi Arabia | 162,400,000 | | |
| Syria | 1,521,000 | | |
| | | | |
| Total | 337,894,000 | | |
| | | | |
| Others | 881,893 | | |
| *Africa* | | | |
| Algeria | 9,440,000 | | |
| Angola-Cabinda | 1,635,000 | | |
| Congo Republic | 1,550,000 | | |
| Egypt | 3,325,000 | | |
| Libya | 21,500,000 | | |
| Nigeria | 16,750,000 | | |
| Tunisia | 1,860,000 | | |
| | | | |
| Total | 56,060,000 | | |
| | | | |
| Others | 1,761,690 | | |

*Source: Oil and Gas Journal* (1982).

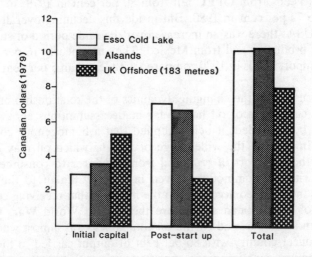

*Fig. 7.1*  Production costs per barrel of oil. *Source*: Browne, E.J.P. 'Canadian in-situ oil sands developments', Institute of Mining and Metallurgy Conference, 'National and international management of mineral resources', London (1980)

the likely average price for all four is considered (Fig. 7.1). Only in terms of *pre-production* expenditure per barrel is the North Sea more expensive.

Thus implicit in the production planning of oil from the tar sands is the assumption that whilst oil will be marketed at prevailing OPEC prices (in 1980 generally in the range $32–6 per barrel but with a maximum of $41 depending on the type of oil involved) it must be produced inside a more favourable national/provincial tax regime than that extant for convention-ally extracted oils. Though this should be negotiated satisfactorily, there is some unease stemming from the imposition in 1980 of tax regimes viewed as unfavourable by oil companies on the output of conventionally produced Alberta oil. This led producers to reduce their output.

Turning to the international picture regarding oil production in the early 1980s (see Fig. 7.3) though OPEC, led by Libya and Kuwait, has as a matter of policy reduced output (in 1980 it fell by 12%), non-OPEC countries have been encouraged to expand output to meet at least some of the difference. In Western Europe, British North Sea output has been increasing modestly (by 3% in 1980). In Africa, Ghana, Cameroun, the Congo and Angola have raised their output sharply (by 6% in 1980), and Mexico responded even more spectacularly with a 36 per cent increase in that same year. These, albeit limited, changes in the levels of oil supply have been reflected in the import patterns of most of the industrial countries. Consumption had been steadily declining towards the end of the 1970s in the USA because of price rises, the economic recession and fuel switching, and the percentage import of total oil needs from OPEC fell from 80 per cent in 1978, to 77 per cent in 1979, to 72 per cent in 1980. But inside this declining overall market for oil in the USA, there was an increase over the same period of supplies from North Sea producers and from Mexico. They met nearly 16 per cent of US crude oil import needs in 1980 compared with less than 5 per cent three years earlier.

The discussion of the changing dynamics of the relationship between suppliers and consumers of oil has so far made assumptions about the nature of what is being traded, it being implied that a homogenous commodity is involved. In spite of the wide range of uses to which oil may be put, the notion of the transport of *crude* oil from producer to consumer is largely accurate, with the commodity moved in bulk by tanker to the entry port where it is then refined according to the needs of that receiving country. But this has not always been so. Before the Second World War, refinery capacity, then as now owned by the oil companies, was almost wholly located in the producer country with 80 per cent of output carried to the consumer in small tankers containing one specific fraction. With consumption before 1939 at comparatively low levels and the number of products relatively small in even the highly industrialised countries, this made sense.

The complete reversal of that situation, however, occurred in the immediate post-war period for a number of reasons, so that by 1980 only 20 per cent of oil output was carried as separate identifiable refined products. First, whilst demand for oil rose steeply in the industrialised countries, the techniques used in refining had progressed sufficiently to enable oil products to be produced which were much more closely tailored to the needs of the local market thus justifying the construction of refineries in the consuming country. Second, technical advances also meant that the amount of unusable waste to be disposed of from a delivery of crude oil was greatly reduced. Third, political uncertainty in producer countries and in particular the experience of the Anglo-Iranian Oil Company (now British Petroleum) which had its refinery at Abadan suddenly nationalised in 1952 without compensation made such operators increasingly wary about the location of such high cost plant. Finally, it was realised that the use of small tankers each carrying a separate refined product was not a cost-effective means of moving oil over long distances compared with the economies of scale which might accrue using large tankers filled with just crude oil. As Fig. 7.2 makes clear, if the cost of moving a tonne of oil in an 80,000 tonne ship is 100, the cost in a ship of 270,000 tonnes will be 50 and in that of a 540,000 tonne ship, only 44. It should be added, however, that at the end of the Second World War, tankers were no more than a quarter of the smallest size represented in this diagram.

Although these common factors determined the relocation of much refining capacity in the consumer countries in the post-war period, there are distinct variations in the nature of that capacity at different locations in the industrialised world. In the USA the capacity rose only slightly over the

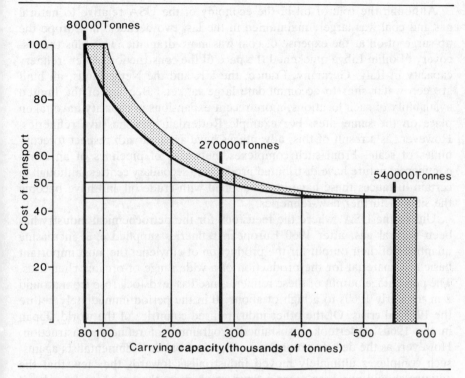

Fig. 7.2  Economies of scale derived from large tankers. *Source*: Shell Trading and Transport Company

1970s as compared with Western Europe and Japan, a situation partly determined by the fact that for much of this century the USA has been self-sufficient in oil. The tradition of small domestic refineries founded on local oil production has therefore been maintained in spite of the fact that it is now an importer of oil. An OECD (1973) report suggested that over half of refining capacity in the USA was concentrated in 12 per cent of the total number of refineries, each averaging a throughput of around 10 million tonnes per year. As the environmental lobby was to some extent successful during the 1960s and early 70s in resisting the building of new large-capacity refineries more in keeping with the USA's current need to import oil, existing small facilities were increasingly unable to cope. The country found itself having to import its shortfall of refined products mainly from refineries reasonably close at hand in the Caribbean and Venezuela. However, with the more recent decline in overall consumption in the USA and the fall in imports together with rising production at home, it seems likely that the situation will be contained and the low level of refinery capacity construction which persisted through the 1960s and 70s of between 3.5 and 4 per cent a year will prove to have been adequate.

Although the role of oil in the economy of the USA relative to natural gas and coal was largely maintained in the last two decades, in Europe the up-surge of oil at the expense of coal was more dramatic. This plus the discovery of oil in Libya quickened the pace of the construction of new refinery capacity in Italy, Germany, France, the UK and the Netherlands, all built at deep water sites to accommodate large tankers. Because of the limited availability of such locations, more recent expansions in capacity have taken place on the same sites. For example Rotterdam now has five refineries. However, as a result of this, advantages have accrued with respect to economies of scale. From such complexes, networks of pipelines of an intra-continental nature have distributed products to secondary centres, although in certain instances these have been supplied with crude oil and have become the sites of further new refineries.

Unlike the USA, where the feedstock for the petrochemical industry has been natural gas, after 1960 European refineries supplied ever increasing quantities of their output for the production of ethylene, the most important basic raw material for the production of a wide range of organic chemicals. The percentage output of these refineries used as feedstock rose from around 2 in the early 1960s to a high of about 10 in the period immediately before the 1973 oil crisis. Of the other industrialised countries of the world, Japan in the 1960s undertook a sustained programme of refinery construction. However, as the decade progressed, pressure from environmentalists against such complexes ultimately moved industrialists towards the view that the country should purchase more refined products at the expense of crude oil particularly from the Middle East. Whilst this policy seemed coincident with the desire expressed by OPEC members during the late 1970s and early 1980s to establish for themselves export refineries and petro-chemical industries so as to maximise the home-based profits to be derived from their indigenous reserves of oil, Japan has subsequently found itself in no sense favoured by Gulf producers. Indeed, when oil supplies were short at the time of the major crises of the Arab–Israeli and Iran–Iraq wars, it found itself as much affected by cutbacks in supply as other countries but in a situation where it had relied for over 73 per cent of its energy on imported oil. Unlike other industrial nations it has no coal or natural gas of its own. Japan has therefore recently opted for a policy of pursuing co-operative production arrangements with non-OPEC exporters in Latin America and in the Far East, especially Indonesia and Australia.

## 7.2 Natural gas

As already noted, the massive increases in the price of oil and restrictions in, and possible interruptions to, its supply have made it necessary for the most vulnerable industrial nations to diversify their energy sources. One way in which this has been achieved in the short term has been to turn to natural gas. Since the conditions under which this resource which mainly consists of methane is formed resemble those for oil, such a move might seem surprising and to present many of the same constraints as oil. The pattern of world reserves established at $78,000 \times 10^{12}$ cubic metres suggests the pre-eminence of the USSR with 41 per cent of these reserves and Iran with 14.5 per cent.

However, there are instances in which the formation of natural gas has been associated with coal deposits. The Dutch gas field of Groningen is thought to be derived from the action of heat and pressure on the coal measures which lie below the Permian gas reservoir. Thus there is a wider spread of gas reserves. Whilst the USA does hold some 7.5 per cent of the world's total, the remainder is much more widely distributed in Western Europe, Australia, the Far East, Africa, South America and Canada. In terms of output in 1980, the USA predominated with 26 per cent of the total, but the USSR extracted 24 per cent, the Netherlands and Canada nearly 5 per cent each and the other areas a collective 30 per cent (Fig. 7.3).

The advantage of developing gas fields compared with oil lies not so much in lower initial capital costs but because much less processing is required prior to the use of the product. Regarding North Sea gas, all that has been required once it is brought ashore is a means by which it may be pumped and de-watered. However, one of its disadvantages regarding international trade is its high cost of transport if this involves liquification. Although such a procedure means a contraction in its volume by a factor of six hundred, it has to be kept at $-160\ °C$ and can only be moved in complex and expensive specialised ships. Thus it costs $0.90 per barrel to transport crude oil, but gas can be the equivalent of $6.50 per barrel. Prices which reflect such high transport costs are currently being paid by Japan for supplies in liquid form shipped from the Middle East and from Indonesia. Thus it is hardly suprising that trade in natural gas by such methods was the energy equivalent of less than 5 per cent of total oil trade in 1980. However, where pipelines are concerned and the gas may be moved at normal temperatures, the costs per kilometre transported are highly competitive. For this reason gas was in 1983 about to be moved under a new agreement from Canada to the United States eastern seaboard states by such methods, and Canada itself was establishing its own pipe network to move supplies from Alberta, where the number of gas wells now exceeds 10,000, to the Atlantic provinces.

In the USA gas makes up 25 per cent of its energy needs. In the UK the figure is around 20 per cent. This will rise once the network for gathering gas from the North Sea fields (including those owned by Norway) is complete in 1986 and is able to tap reserves amounting to $0.564 \times 10^{12}$ cubic metres compared with $0.15 \times 10^{12}$ cubic metres at present. In Western Europe, 1980 saw the completion of a pipeline which connects with all countries enabling 17 per cent of their total energy needs to be met by natural gas. If discussions begun in 1983 to tie this into a pipeline system tapping the huge Siberian reserves of the USSR come to fruition, Western European energy dependence on gas could rise dramatically to the ex-

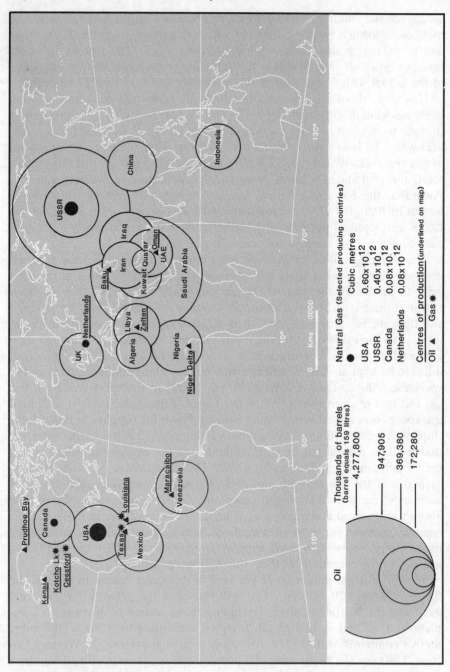

*Fig. 7.3* Oil, natural gas: output by country with chief centres of production, 1980. *Source: Mining Annual Review, Minerals Year Book*

tent that it could become the major alternative energy source to oil. The strategic desirability of such a move, however, involving great reliance on a single supplier of alien political complexion, would be open to question. Although ultimately much of Europe could become dependent on coal imports in the medium term were coal to become the chief alternative to oil, at least these might be brought in from a number of nations all operating within a more congenial socio-economic framework.

## 7.3 Coal

Any appraisal of the development of the industrial economies following the Second World War must recognise their increasing reliance, certainly up to 1973, on oil, as if the aggressive marketing policies of the oil companies could of themselves ensure in perpetuity the continuing availability of this commodity in whatever quantities were required but at a consistently low price. Although in retrospect such a view would appear to have been naive, the fact that immediately before the crises of the 1970s oil made up roughly half the energy consumption of the world, yet represented less than 40 per cent of known energy raw materials compared with coal which represented some 60 per cent of reserves, ought to have provided grounds for serious thought. However the reality of the situation following the actions of OPEC was fortunately realised and not surprisingly coal is now being looked to as the best means of supplying an increasing amount of our energy needs.

A report of the World Coal Study entitled 'Coal – Bridge to the Future' (1980) suggested that the contribution of coal to energy needs should be increased from the current one-quarter to well over a half by the end of the century. To achieve such an end world production must rise by 2.5 to 3 times, whilst the world trade in steam coal will have to grow 10 to 15 times 1980 levels. However the experts who produced the World Coal Study were confident of the role it can play in the short to medium term, acting as a bridge between oil and the energy systems of the future. They predicted that coal can remain competitive, particularly in the field of electricity generation with technological advances in combustion, gasification and liquefaction greatly widening the scope for the environmentally acceptable use of coal. This is particularly necessary when it is realised that conventional coal burning power stations, especially in Western Europe and the eastern states of the USA, have been and remain major sources of sulphur dioxide ($SO_2$) and to a lesser extent, nitrogen oxides ($NO_x$). Such emissions, if released into the atmosphere as a result of the direct burning of coal in a conventional mode, may contribute to acid rain, a phenomena which may have an adverse impact both on vegetation and fresh water bodies well beyond the boundaries of the nations initially responsible for this form of pollution. Whilst some have supported such contentions (for example, The Royal Society of London, 1973) and have suggested that the discharge of such gases from

UK coal burning power stations has, through acid rain, adversely affected lakes in Scandinavia, this is not a universally held view. The UK Central Electricity Generating Board maintain that a case has yet to be made and that the removal of these emissions from its effluents would cost in excess of £4,000 million (1983 prices). However in an effort to establish the true situation the Central Electricity Generating Board in collaboration with the National Coal Board of the UK agreed in September 1983 to sponsor a major international research project. Under the direction of the Royal Society it will last five years, cost £5 million and involve scientists from Norway and Sweden as well as the UK.

In taking an immediately more pragmatic view the International Energy Agencies Coal Industry Advisory Board have criticised the lack of urgency with which some countries have been responding to the worsening energy situation. Their commitment to increase coal usage has seemed much slower than is necessary, particularly where electricity generation is concerned. The Board recommended that in OECD countries coal-fired capacity should be raised from 350 GWe to 1,100 GWe by the year 2000. It suggested the further construction of new oil-fired power stations should be banned, oil-to-coal conversion encouraged and financial and technical aid be made available to less developed countries to develop coal production and use. The extent to which countries accept these recommendations now emanating from the Board and from the other study is a matter of conjecture. What does seem beyond doubt is that a renaissance for coal is imminent and the next twenty years will witness a growth rate in its production and consumption not to be paralleled in any other sector of the minerals industry.

The truth of such a statement is already beginning to become apparent in the USA, the country with the largest annual output of coal though it has less than 18 per cent of the world's reserves (Fig. 7.4). Despite a fall in industrial activity, production rose there by 7 per cent in 1980 bringing total output to 830 million tonnes. Both export and domestic consumption increased by more than 8 per cent. Whilst these trends were certainly a response to rising oil prices and possibly to set-backs experienced by the US nuclear power industry and its ultimate capacity to deliver much in the way of energy supplies over the next few years, the fact remains that coal only provides 20 per cent of energy supplies. Under these circumstances the President's Commission on Coal has recommended policies that will double production by 1990 to over 1,500 million tonnes a year with most of the additional coal (768 million tonnes) coming from the western states. By that date the aim is to have coal accounting for at least 28 per cent of total energy with the transfer from oil to coal burning power stations supported through the period by federal aid. Though it is expected that such conversions will cost about $15,000 millions they will ultimately save oil consumption of the order of 200 million barrels a day.

Since most of the new US mines will be underground the current split between surface and underground mining for coal which is 60:40 is expected

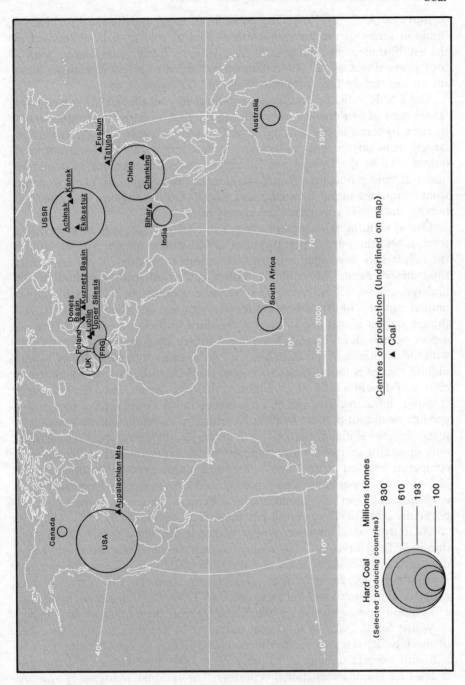

*Fig. 7.4* Coal: output by country with chief centres of production, 1980. *Source: Mining Annual Review, Minerals Year Book*

to move to 45:55 by 1990. The possibilities suggested by the World Coal Study in terms of the less conventional uses of coal have been realised in the establishment by Congress of the Synthetic Fuels Corporation. With a four year budget of $20,000 millions they intend to invest directly in plant set up to produce fuel from coal shale and tar sands.

The USSR is the second largest producer of coal though it has well over 60 per cent of world coal reserves. Since the first oil crisis, output has steadily risen by some 14 per cent to 715 million tonnes a year with the expansion largely from surface workings which are now responsible for a third of total output. Whilst the Donets and Kuznetsk basins will continue to dominate under-ground production, future increases in coal output are likely to come from complexes in the Ekibastuz and Kansk-Achinsk regions where surface mining and power generation are closely integrated.

China, with the third largest reserves of coal at 17 per cent of the world's total, is also the third largest producer at 610 million tonnes in 1980. Unlike the industrially developed countries, this mineral provides 70 per cent of its total energy needs. Most of the output is derived from a multiplicity of underground workings. Even so twelve large coal mining areas each with an annual capacity of 10 million tonnes per year are the main producers. Although output is planned to rise by one-third by 1990, the most interesting aspect of new developments are plans to open up a number of new fields with the assistance of the Japanese. One field with a total potential of 11,000 million tonnes is in the Junga section of the Mongol region whilst a further eight could yield a total of 23 million tonnes per annum, once the necessary transport infrastructures can be established. Japan would receive part of the productive output of these fields as its return on investment of cash and technical expertise. Poland, ranked fourth among world producers, has an output only one-third that of China and reserves one-tenth the size. Nevertheless, compared with other producers it is still a major exporter sending abroad in 1980 over 20 per cent of its production. Although output from the major coalfields of Upper Silesia and Lublin is expected to grow, internal political problems seem likely to curtail any immediate expansion.

UK output at 127 million tonnes makes it the world's fifth largest producer. With reserves at 7,000 million tonnes including the two most recent discoveries, the fields of Selby and the Vale of Belvoir, it has sufficient coal for another 50 years at current rates of extraction. By the year 2000 it is expected that output will have expanded to 170 million tonnes per year with 150 million tonnes from deep mines and the rest from opencast workings.

Whilst South Africa's total coal output is 14 per cent smaller than that of the UK, its reserves are fourteen times greater. Output is rising rapidly and with exports running at 20 per cent in 1980 it is likely to become one of the four major international exporters. The International Energy Agency predicts that exports will reach 94 million tonnes by the year 2000 when eight major new projects will have added more than 70 million tonnes to annual capacity. By that date coal as a money earner will rival, if not surpass, that of gold.

Although it has slightly smaller reserves than South Africa and a marginally lower output, Australia exports more than double the current levels of that country. Over the next twenty years exports are expected by the *Mining Annual Review* to increase by a factor of five from a range of new mining developments and port facilities being developed in Queensland (Mackay and Gladstone) and New South Wales (Newcastle and Port Kembla). This large increase is in response to a growing international demand for steam coal especially from Japan which already takes nearly 70 per cent of all coal exports. By the year 2000 Australia will become second only to the USA as an exporter of coal. Unlike the USA, by 1990 coal is likely to become the major source of energy in Australia.

The German Federal Republic, like the UK, faced a situation of steady decline in its output of hard coal through the 1960s and 70s in the face of competition from cheap oil and natural gas. From 1950 to 1973 production declined from 150 million tonnes to 85 million tonnes and 75 per cent of the pits were closed. In the wake of the oil crises, the industry, largely centred on the Ruhr coalfield, was reorganised and modernised and entered into contracts with the electricity industry that should ensure its viability to 1995. Output may reach 100 million tonnes by the end of the century. This figure is likely to be 30 to 40 million tonnes below expected overall demand. Its future beyond the year 2000 may largely depend on the ability to work coal at increasing depths and the eventual validation of promising deposits in the Lippe valley. In contrast to the problems which have beset the hard coal industry, the output of the opencast brown coal in the Rhineland district was maintained through the era of cheap oil. This was almost certainly the result of its low cost and its close association with local power stations. It currently produces around 131 million tonnes per year and with deposits extending over 2,500 km$^2$ its future seems assured.

Perhaps the worst record of all the main coal producers in terms of its contribution to domestic energy supplies since 1973 is Canada. Although the current figure is 10 per cent this is only expected to rise to 18 per cent by the end of the century. Like Australia, many of its newest coalfield developments have been undertaken with the Japanese market in mind, particularly those of Alberta and British Columbia. Of the 13.7 million tonnes of thermal and coking coal available for export in 1980, that country took 78 per cent and future expansion in British Columbia as well as Alberta is expected to further underpin the Japanese steel industry. To this end new port facilities are being developed at Prince Rupert to supplement the current terminal south of Vancouver. By the end of the century exports of metallurgical coal are expected to reach 23 million tonnes per year and exports of steam coal to be in the range 24–44 million tonnes per year, though some of the output of the latter will certainly be required for steam raising to increase liquid production yields in the oil sands developments in Alberta.

Of the three countries (Canada, Australia and South Africa), with the greatest immediate development potential in terms of coal outside the planned economies, the World Bank has estimated that investment of the order of $60 billions will be required between 1980 and 2000 if their plans are to be realised. For the Canadian and Australian enterprises the political stability of these countries would mean that although the sums involved are large, international finance could be found. With considerable profits at their disposal due to increases in the price of crude oil held as stock, the oil companies have been diversifying their interests in an effort to become comprehensive providers of energy. It is from these sources that the huge capital sums will be provided. In the USA alone well over a third of the coal industry is now controlled by the oil companies. However what is more immediately evident is that these countries are not only industrialised nations in their own right, but will have increasingly large export surpluses which they will be willing to sell to other developed countries. Japan has already been cited as a customer of all three whilst Canada is developing market ties with both Western and Eastern Europe. For this reason the other industrial countries of the world that may be net importers of coal are unlikely to find themselves, once they have come to rely on coal as their main source of energy, vulnerable in the way that they were with oil after the 1973 crisis. Nevertheless, a number of the industrial countries of Western Europe could, in the medium term, become as dependent on imported coal as they are now on imported oil.

**CHAPTER 8**

# Minerals and the environment

## 8.1 The environmental impact of mineral extraction: underground vs. surface operations

The magnitude of some modern open-pit operations has already been illustrated by the Lee Moor china clay workings which contribute some 20 per cent of British output (Ch. 1). Similar but much larger scale operations in the St Austell area of Cornwall provide almost the whole of the other 80 per cent, so that the environmental impact of this particular extractive activity is more concentrated in the UK than for any other mineral. Unfortunately as Chapter 1 also indicated, open-pit activities are not the only agent of landscape modification since massive tips of quartz waste characterise the adjacent areas. The extent of the present working areas and those that will be required later for both extractive operations and waste tipping in the St Austell area indicate the magnitude of the problem (Table 8.1). Indeed the land requirements for the disposal of all non-saleable products (at about

*Table 8.1*   Land requirements (in hectares) for the china clay industry in the St Austell area

|  | 1970 | 1980 | Probable requirements up to 2029 |
|---|---|---|---|
| Pits | 646 | 820 | 2491 |
| Tips, (sand, stent, overburden disposal) | 830 | 1112 | 3776 |
| Micaceous residue disposal | 77 | 198 | 146* |
| Plant areas | 310 | 370 | 540 |
| Total | 1862 | 2500 | 6953 |

* Assumes some use of disused pits for disposal of micaceous wastes.

*Source*: China Clay Development Association.

*Fig. 8.1* Bingham Canyon Copper Mine, Utah, 48 km south-west of Salt Lake City. *Note*: The pit is 4 km wide, about 1 km deep and is reputedly the largest man-made excavation in the world. Output is running at about 110,000 tonnes of ore per day. *Source*: Kennecott Copper Corporation

1,310 hectares) exceeds that of the pits (820 hectares). This is hardly surprising since the ratio of waste to valuable mineral is about seven to one in the china clay extraction industry. In 1980, the total output of 2.8 million tonnes of refined clay meant that nearly 20 million tonnes of waste was surface tipped.

Excluding the rather exceptional problems of waste, the increasing impact of china clay working is not wholly untypical of most other minerals both in the UK context and the world at large. As far as the former is concerned, apart from oil, lead, tin and salt, all other minerals showing a substantial increase in output over the period 1950 to 1980 (between over 22% for chalk to nearly 300% for sandstone) were worked by open pit or quarry. The environmental change wrought by the working of aggregate materials, that is materials with the largest percentage increases in 1980 over 1950 (as Table

8.2 makes clear) is likely to become even greater in the period up to the year 2000 once demand for construction material begins to rise again after the recession of the 1970s and early 1980s. Then aggregate producers will probably look for further economies of scale, though these may be achieved by further concentration at those sites where the extent and quality of the reserves offer the best exploitation opportunities in relation to markets.

In the UK greater intensity of activity regarding hard rock materials is apparent in such areas as the Mendip region of Somerset, the Charnwood Forest district of Leicestershire, the Craven District of North Yorkshire and the Brecon Beacons in South Wales. As for sand and gravel, since deposits have tended to be found in areas close to urban development, such a solution is less satisfactory. For instance the McLellan *et al.* (1979) study of the Waterloo area of Southern Ontario showed that apart from urban development and agriculture, sand and gravel workings were the most significant land use at 2,707 ha. In London's Green Belt alone 3,320 ha have been actively worked for minerals, almost wholly sand and gravel, and a further 3040 ha have been approved for the same purpose. Nearly all of this is localised in areas with significant pressures from other incompatible land uses (Standing Conference 1976). These examples must reinforce the view that sand and gravel operations are not only extensive in such areas but, because of their close proximity to urban activity, are a significant potential source of environmental disruption through general unsightliness and noise. Public dissatisfaction is inevitably all the greater and more vociferous where such operations are also associated with programmes of further processing. A survey carried out in the UK during the mid-1970s indicated that local authorities receive more complaints about sand and gravel workings than any other extractive activity even allowing for their ubiquity (Blunden 1975). Although there were initially some grounds for believing that such environmental problems might be ameliorated by increasing the output of sea won materials by dredging (a common practice off the coasts of south-east England and the Eastern seaboard of the USA), claims have been made about their interference with navigation, submarine cables and pipelines. Even more importantly, they allegedly damage fishing grounds and cause coastal erosion.

Land won aggregates have always been worked by open-pit methods but taking a broad view of mineral extraction on a world-wide basis there has in this century been a marked swing away from underground mining towards surface methods of working. This has been especially true in the United States where, until the mid 1960s, underground mineral working predominated. By 1980, however, over 80 per cent of that country's mineral wealth was extracted by open-pit methods. This trend is to some extent due to the fact that, as more disseminated metalliferous mineral ores of lower mineral content have to be won, such techniques are more appropriate. It is also due to a widening gap in the respective productivity of open-pit methods compared with underground extraction. The former technique has been favoured by modern mechanised earth-moving equipment. In the 1960s it was esti-

Minerals and the environment

*Table 8.2*   Production of minerals* in the UK: 1950, 1975 and 1980 (thousands of tonnes)

| Mineral | 1950 | 1975 | 1980 |
|---|---|---|---|
| Coal | | | |
|   deep mined | 207,400 | 117,400 | 112,400 |
|   opencast | 121,100 | 10,400 | 15,800 |
| Natural gas† | | | |
|   North Sea | — | 53,944 | 55,008 |
| Crude oil† | | | |
|   North Sea | — | 1,223 | 78,917 |
| Iron ore | 13,170 | 4,490 | 916 |
| Non-ferrous ores | | | |
| (metal content) | | | |
|   tin | 1.4 | 1.1 | 3.3 |
|   tungsten | 0.1 | — | — |
|   lead | 4.3 | 6.4 | 3.6 |
|   zinc | 0.1 | 4.0 | 4.4 |
| China clay } | | | |
| Ball clay } | 1,131 | 3,220 | 3,964 |
| Clay and shale | 23,956 | 27,794 | 19,825 |
| Slate | 150 | 547 | 225 |
| Limestone | 25,365 | 92,898 | 74,713 |
| Dolomite | 1,400. | 2,315 | 14,060 |
| Chalk | 13,138 | 17,924 | 14,049 |
| Sandstone | 4,150 | 13,352 | 12,597 |
| Sand and gravel | | | |
|   land | 38,000 | 110,028 | 87,431 |
|   marine | — | 14,000 | 17,036 |
| Igneous rock | | | |
| (mainly granite) | 11,423 | 42,017 | 34,676 |
| Gypsum | 1,181 | 3,258 | 3,371 |
| Rock salt | 42 | 754 | 1,746 |
| Salt from brine | 985 | 1,740 | 1,608 |
| Salt in brine | 3,286 | 5,136 | 3,800 |
| Fluorspar | 51 | 235 | 186 |
| Barytes | 81 | 52 | 54 |
| Potash | — | 26 | 509 |

* Major minerals only          † Approximate coal equivalent

*Source*: Natural Environmental Research Council (1982).

*Note*: Apart from the rapid growth in the aggregates market in the period 1950–75 this table highlights a number of trends also applicable to Western developed countries as a whole during such a period of sustained economic expansion. The increasing problems of working deposits of sand and gravel which tend, when land based, to be located in close proximity to urban areas, were countered not only by significant increases in sea won sand and gravel, but also in rises by a factor of three in the output of alternative aggregate materials such as sandstone and igneous rocks. Notable falls in coal output are also discernible in this period as the economies of the UK and other developed nations switched to oil and natural gas as their main mated that the surface extraction of minerals could achieve a productivity rate of 500 tonnes per manshift. In contrast, a productivity rate from underground mining of 50 tonnes per manshift was common. Whilst such trends continue the take of land for open pit working is bound to grow at a high rate.

The rate of environmental change wrought by this situation will only be partially offset, where aggregate materials are concerned, as the balance of aggregate production moves in favour of hard rocks. This is because materials such as granite are relatively more economic in terms of land consumption compared with sand and gravel.

However, the inevitable increase in the demands made on land year by year will be further intensified as large-scale open-pit workings for non-ferrous metals involve the extraction of lower grades of ore. Not only will larger operations be needed to produce the output of metal required but the leanness of the grades of ores will need to be offset by the massive scale of open-pit operations and by high levels of output per annum if costs per tonne extracted are to be minimised. Such operations are not always going to be of the order of Bingham Canyon open-pit copper mine in Utah (Fig. 8.1), but nonetheless they could have considerable environmental impact. The hypothetical proposals for a working copper mine in a highland area of Britain illustrate this point well enough (Commission on Mining and the Environment 1972). Fig. 8.2 indicates how the working area of 325 hectares, within the site of 800 hectares, would contain a pit of some 81 hectares. This underlines the need, wherever open-pit metal mining is concerned, for large areas to be given over to the disposal of waste rock overlying the ore (that is overburden) and the waste materials resulting from minerals processing (that is tailings).

## 8.2 Surface waste disposal and environmental change

Although the tipping of overburden is not an agent of environmental change where underground working is involved, such operations frequently involve the surface disposal of waste rock and/or tailings. For deep mined coal in the UK, the average ratio of waste to saleable product in the 1970s was 1:2 and a total of between 61 and 71 million tonnes of waste was tipped annually. This ratio is, of course, substantially better than that of china clay

---

sources of energy. But the 1950 figures are compared both with 1975 *and* 1980 to provide some measure of the economic decline of the last half of the last decade. Falls in the amount of aggregate materials used after 1975 show the cut-back in the building industry, always an immediate and responsive measure of national economic health. The almost static level of coal output in the period 1975–80 is the first sign of the revival in the market for coal following the massive rises in oil costs during the last decade.

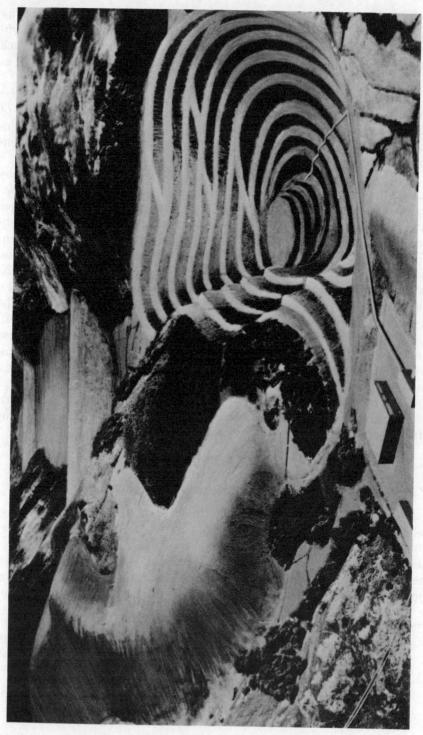

and that of the now largely defunct UK slate industry where waste could rise to twenty times that of the commercial product. However, the problem is substantially greater with coal because of the much higher tonnage of output produced, the potentially more volatile nature of the shale waste and the greater number of production sites compared with both china clay and slate.

Whilst attempts have been made to use such materials (see Ch. 9), the problem in terms of environmental change remains the most outstandingly difficult of all those extractive operations involving waste disposal. In the UK county of Nottinghamshire alone over 900 hectares of the country are covered in spoil heaps, with the collieries disposing of about 13 million tonnes per year. The local government authority of the county estimated that using present tipping methods and assuming production constant at 1980 levels, 5,000 hectares of the county willl be covered by waste tips in the year 2,000. However, there as elsewhere, increasingly stringent standards of coal preparation resulting from more sophisticated customer requirements are resulting in a proportionate increase in waste production. At the same time the decrease in discrimination between coal and waste which results from mechanised mining is also having this effect and fully mechanised pits may have waste to coal ratios approaching 1:1. All these factors will be further exacerbated as demand rises.

Underground metalliferous mining operations, like those of their open-pit counterparts, also necessitate the removal of the metallic ore with the parent rock as it is dug from the ground, however selective the mining operation may be in following the vein. Thus tailings disposal is again a problem. These are usually disposed of in a nearby artificial reservoir (Fig. 8.2). Such areas usually can be fitted into the floor of an adjacent valley without serious damage to the landscape, but problems arise when they are completed. Although they can be de-watered, their surfaces often present a hostile environment to plant life and thus make the ready establishment and maintenance of a vegetational cover difficult. Moreover, the water draining from these areas may be extremely acid (where the working of sulphide or pyrrhotite mineral ores has been undertaken) or contain toxic metal ions. In both situations, the polluted water, once introduced into the natural drainage system of the area can cause widespread damage by destroying all aquatic life. This has proved an environmental problem in many areas, especially in North America where mining operations are on a large scale and have, in the past, been relatively uncontrolled in environmental terms. In the Sudbury region of Ontario (Canada), for example, it was found that nickel levels in a river draining the area were 35–55 parts per billion compared with 5–7 parts per billion for rivers which did not drain metal contaminated areas (Stokes *et*

*Fig. 8.2*  A hypothetical open pit copper mine in a highland area of Britain. *Note*: overburden is tipped to the left of the open pit. To the centre rear of the picture the tailings area can be seen behind the dam wall. The processing plant is in the foreground. *Source*: © Commission on Mining and the Environment. (Photo: Rio Tinto Zinc.)

*al*. 1973). The working of uranium ores around Eliot Lake, Ontario, is another case in point, where the strategic need for the material was such that little was originally done to contain the entry of acid tailings water into the meandering drainage system which exists on the edge of the Laurentian Shield. The result has been widespread environmental devastation.

Areas used for tailings deposition, once dried out, can also offer another environmental hazard if their surfaces are left exposed: that of dust blow. In Sudbury the blow of fine material from the extensive areas of copper and nickel tailings has proved a less than pleasant experience for the adjacent local communities. A similar problem deriving from the uranium tailings in the Eliot Lake area posed both inconvenience and danger from radio-activity for people and crops (Ripley *et al*. 1978). Although all mining operations produce some dust, where aggregate materials are extracted the problems are usually quite small and localised irrespective of the size of the operation. However, where such enterprises are close to urban areas, nuisance can be caused. In this respect plant processing limestone or chalk for the production of cement have produced most of the relatively infrequent problems.

## 8.3 Smelting and environmental pollution

Of far more consequence with wider, if not regional implications are those instances where metalliferous mining operations have also led to on-site smelting of ores. Problems derive not only from the emission of particulates but especially of sulphur dioxide ($SO_2$). In the lead belt of Missouri, USA, the local smelter near Rolla which emits 75 tonnes of $SO_2$ and 200 tonnes of lead particulates every day has damaged vegetation in the surrounding Clark National Forest to the extent that the whole area is now being carefully monitored (Wixson 1972). However, the uncontrolled release of $SO_2$ in earlier years of the century from open air smelting operations at Sudbury based on the local mining of nickel and copper, and at Trail, British Columbia (Canada), related to the mining of lead and zinc, have had an even more devastating impact on vegetation. In the Trail district the average output of 300 tonnes of $SO_2$ per day caused the total destruction of many thousands of hectares of vegetation in the valley in which the smelter was located, the $SO_2$ frequently being trapped by temperature inversions. Since 1929, when a process of utilising much of the $SO_2$ that would have been released into the atmosphere in the manufacture of sulphuric acid was introduced (thus permitting a reduction in pollution of around 95%), the whole area has been revegetated (Blunden *et al*. 1973). However, the situation at Trail was made more complex by the close proximity of the US border. The result was that, coincident with the taking of remedial action around Trail, an International Joint Commission of Inquiry found that the smelter had been responsible for damage to the vegetation inside US territory making the polluting company liable for its actions across the national boundary (Sandbach 1982).

At Sudbury the situation with regard to damage was considerably worse and far more widespread with much higher levels of $SO_2$ discharge. There was much subsequently done to contain the problem by the simple expedient of undertaking smelting in purpose-built plant (instead of in the open air) with effluent dispersal achieved by chimneys. But with rising production levels, it was not until the late 1960s that substantial reductions in $SO_2$ of 40 per cent were achieved by using the gas to make sulphuric acid. However tree growth was still affected and there was evidence of $SO_2$ damage to Eastern White Pine trees and other sensitive species (Linzon 1972), particularly in areas immediately south-west of Sudbury such as the Whitefish Lake Indian Reserve where $SO_2$ had been identified at between 0.01 and 0.02 parts per million (Fig. 8.3). This persisted until 1972 when the key INCO smelter was equipped with a 387.5 metre stack designed to achieve a high level of gaseous effluent dispersal well away from the Sudbury area. Now the stack appears to have exported what were local pollution problems to areas as distant as 50 km down wind of the smelter. In a zone where it seems the material contained in what is still a relatively concentrated plume tends to fall back to ground level as acid rain (the $SO_2$ having been converted by the action of the moisture in the atmosphere to dilute sulphuric acid), the large water bodies of the Killarney Provincial Park area have had the chemical balance of their waters altered (Fig. 8.4). Being in their natural state close to the margins of acidity and lacking any form of alkaline buffering from their granitic bedrock, acid rain in particular appears to have tipped the pH balance of many of these lakes to values below 5.5 providing an environment in which fish and the micro-organisms on which they live cannot survive (Craig 1974). Moreover, the same water bodies have also been found to contain abnormally high levels of heavy metals, particularly nickel and copper (see Figs 8.5 and 8.6) which must have been discharged from the plume into the water bodies as particulates (Hutchinson *et al* 1975). Although there is no evidence here, because these lakes are biologically dead, it is known that metals of this sort can enter aquatic ecosystems and become concentrated in animal food chains to devastating effect. For copper and nickel, evidence shows that if they are taken into the food chain at the same time they can act synergistically, each enhancing the other's toxicity (Hutchinson 1973).

Smelting operations and other forms of secondary metal processing are also known to have had an adverse impact on man. Both mercury and cadmium have been mentioned in this context in Chapter 6, although in the case of the former problems have only arisen due to the discharge of mercuric compounds into water bodies where they have been converted through bacterial action to organic mercury and taken into aquatic food chains. Though the results for man in eating fish in which the mercuric compounds became concentrated were dire enough – in the best documented case at Minimata Bay in Western Kyushu, Japan, 121 people were affected of whom 46 subsequently died (Blunden 1977) – cadmium has proved a health hazard in a

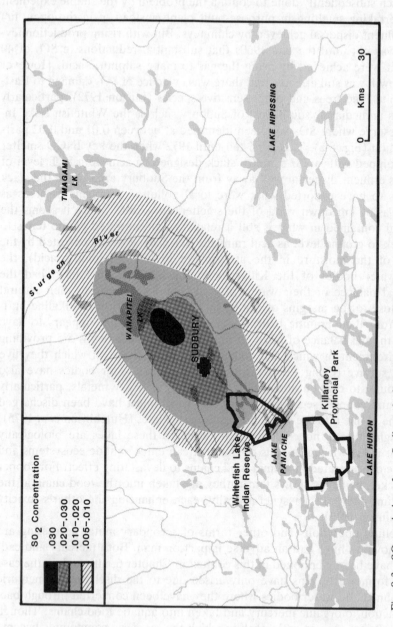

*Fig. 8.3* SO₂ emissions from Sudbury before the construction of the 387.5 metre stack at INCO. Source: Adapted from Beamish, Van Loon, Macfarland, Lichwa (1976)

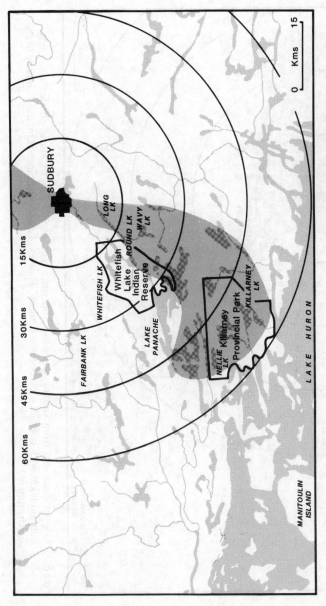

*Fig. 8.4* Emissions from Sudbury after the construction of the 387.5 metre stack. Shaded areas may contain lakes with pH 5.5 or less, depending on the bedrock. *Source:* Conroy, Hawley, Keller, LaFrance (Ontario Ministry of the Environment)

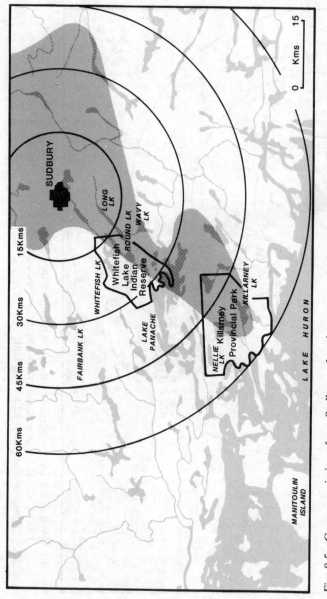

*Fig. 8.5* Copper emissions from Sudbury after the construction of the 387.5 metre stack. Shaded areas may contain lakes with over 25 μg/l. *Source:* Conroy, Hawley, Leller, LaFrance (Ontario Ministry of the Environment)

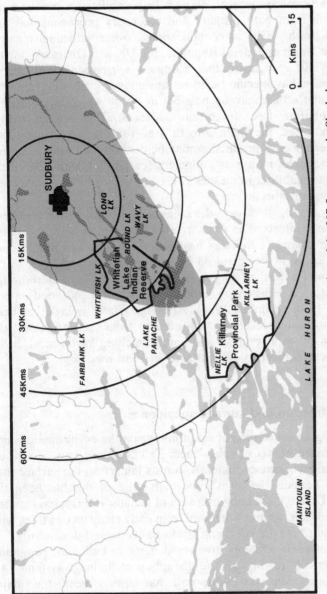

*Fig. 8.6* Nickel emissions from Sudbury after the construction of the 387.5 metre stack. Shaded areas may contain lakes with over 30 µg/l.

*Note:* Nickel can be used as a reasonable definer of metal emissions from the Sudbury area smelting complex since it is naturally low in lake waters. A global mean is 10 µg/l, whilst lakes in the study area can be as high as 300 µg/l, even though there are no direct discharges to these lakes. *Source:* Conway, Hawley, Keller, LaFrance (Ontario Ministry of the Environment) more direct way. Like mercury poisoning, it was recognised first in Japan manifesting itself in an outbreak of a disease called Itai-Itai. This has resulted in the discontinuance of the use of cadmium in that sector of Japanese industry particularly concerned with plating. Its potential as a serious water pollutant resulting from both primary and secondary processing has also been recognised by the EEC in general and by other individual member states such as the German Federal Republic, the UK and Denmark, as well as in Sweden. Whilst recognising that cadmium is irreplaceable in certain well defined uses or has specific technical advantages over other minerals elsewhere, its manufacture may be possible in a way which is industrially hygienic and pollution free in certain instances. This applies to the production of batteries, the main growth area for the use of cadmium. However, in other areas its production may be discontinued simply because substitutes are available; its use as a pigment (currently taking 25% of the world market), as a stabiliser in plastics (10%), as well as its application as a plating metallic are all cases in point.

Finally, the processing of clays in the manufacture of bricks has also been shown to have toxic effects. The discharge into the atmosphere of fluoride as a result of the baking of the clay has caused fluorosis (a form of bone degeneration) in cattle in 19 out of 43 farms close by existing brick works in Bedfordshire and a further 20 out of 38 around similar Peterborough works according to UK Ministry of Agriculture, Fisheries and Food reports published in the 1970s (Blunden 1981). Although relatively little has been done in Britain to reduce levels of such effluent from such processes, the fluorine gas given off as a result of aluminium smelting has in most developed countries been subject to effluent recovery and recycling.

## 8.4 Mineral extraction and surface subsidence

The one remaining aspect of the impact of mining on the environment which needs to be discussed is that of subsidence. This is caused only as a result of the underground working of minerals, but its impact on the surface may be considerable. Perhaps the most important cause of this has been the working of coal. Unlike the mining of a vein of metallic ore this has resulted in the removal of whole seams of material frequently many metres thick with the inevitable result that the overlying strata has collapsed downwards.

Coal mining subsidence mainly involves damage to buildings, communications and to agriculture through the disruption of drainage systems. Although this, along with waste disposal, has always been the major environmental impact of coal mining and there are many examples of surface blight caused by it, the worst effects of subsidence can still only be controlled rather than totally mitigated. Total avoidance of subsidence would involve the leaving of unacceptably large quantities of coal *in situ*, destroying economic viability. Thus the great advances in the understanding of subsidence since the late 1940s do not relate to prevention but to predicting the degree of physical damage on the surface so that surface development and coal exploitation may be co-ordinated.

The three most important factors governing the extent of subsidence may be identified as: (i) ground movements which are related to the dimensions of the mineral body extracted and can reasonably be predicted; (ii) the general stability of the site before extraction commences; and (iii) the tolerance of surface buildings and so on to the stresses caused by ground movements. Using these factors in association with specific data, a reasonable assessment of the probable effect of mining subsidence is possible and careful liaison between local authorities and the mining agency can obviate damage to surface structures.

Two UK examples of co-operation between local authorities and the National Coal Board to effect planned subsidence are worth noting. When the plans for Peterlee, a new town in the county of Durham, were first prepared in 1947, about 30 million tonnes of coal remained under its 948 hectares. By taking the percaution of carefully aligning surface constructions, it was found possible to build on land below which as many as three seams of coal remained to be worked. Similar phasing of coal extraction with motorway development took place when that section of the M1 which crosses the Nottinghamshire/Derbyshire coalfield was constructed. In places the motorway profile was built to allow for subsidence (Blunden 1975).

The situation with regard to salt mining compares very unfavourably with that of coal. A method is in use which ensures the removal of salt in solution through the drilling of bore holes, the pumping in of water and the extraction of the resulting brine leaving behind large caves but sufficient salt to support the overlying strata. This process, invented by ICI, is known as controlled brine pumping and largely eliminates damage from subsidence. However, salt is still won by the random pumping from beneath the earth's surface of salt which is naturally in solution because of its interaction with ground water. In the UK this completely uncontrolled method of extraction has caused subsidence which has resulted in the flooding of agricultural land in Cheshire. It also caused severe damage to the urban fabric of Stafford where in 1969 court action was needed to restrain further natural brine pumping activity. The problem, however, remains in Cheshire where the unpredictability of subsidence presents the gravest difficulty since the course of the natural brine runs is not clearly established and the extent of the area from which the solution is derived is unknown. A county survey on the subject produced by the local authority states that 'the risk of gradual settlement as a result of natural brine extraction remains a question mark over many areas in mid-Cheshire'. The same document also showed the scale of the problem. This indicated that by the mid 1970s the total subsidence resulting from natural brine extraction would be equivalent to a drop of 0–5 metres over 119 hectares. In reality, falls of greater magnitude, but varying widely, were experienced over a much greater area (Blunden 1975).

The most notable cases of subsidence caused by the extraction of crude oil and natural gas have occurred near sea-level where flooding could have been the highly unfortunate result. Evidence of this problem has come from an oil field on the shores of Lake Maracaibo (Venezuela), from coastal locations in Japan and the USA, and from the Po delta in Italy (Poland, Davis 1969). Though they will have occurred elsewhere, the depression of the land surface will very likely have gone unnoticed. Subsidence of this kind does not derive from the actual subterranean removal of strata as with coal and salt, but from the reduction in underground fluid pressure in the oil and gas zones. This allows the strata above to settle at a lower level.

Because of its coastal location the subsidence at Long Beach (Los Angeles), California, an industrialised township lying two to three metres above sea level, represents a well documented case study. As a result of the removal from 1936 onwards of oil from the Wilmington field as well as a subsidiary field at Signal Hill, a substantial part of the central urban area began to sink. By the time the overall fall had reached 8 metres it was decided to take action to prevent the area being inundated by the sea. Retaining walls and levees were constructed whilst repairs to damaged oil well bore holes caused by earth movements had to be carried out. The outlay on these projects amounted to over $100 million before attempts were made in the late 1950s to repressure the oil zone by injecting water. This resulted in a fifteen per cent diminution in the subsidence (Place 1972).

## 8.5  Mineral extraction and environmental and planning legislation

The interaction between mining and the environment as it has been described above does not, however, occur in a milieu of complete unbridled freedom where the mining company is allowed to pursue its objectives with disregard for all other considerations. Apart from being constrained by regulations which pertain to the safe working of pits and underground mines, the extractive industries have found themselves increasingly working within a framework of controls relating to the use of land and its environs which are legally enforced and administered by one or more government departments operating nationally or regionally. That this must be so in the UK will have been made apparent in Chapter 1 concerning Lee Moor. It will have been reinforced in a wider context in the discussion about the progressive reduction of air pollutants from mines and their ancillary processing operations, something which would hardly have occurred as a result of the altruistic motives of mining companies. Moreover it was observed in Chapter 4 on the short term supply and demand for minerals that mining companies have been forced to undertake environmental controls, a factor which they plainly view as another cost element but one which may vary spatially in terms of location and the nature of the mineral that is being extracted.

The reason for the development of such controls in most developed countries lies in the view that in highly populated and complex societies where land is increasingly seen as a scarce resource for which at any one time there may be competing uses, the market place cannot be relied upon in terms of equity to decide what activity is carried out there, when, and to whose benefit (though, of course in the case of minerals they can only be extracted where they exist and in conditions which enable them to be termed reserves). At the same time, in the wider general interest the externalities of visual damage and pollution cannot unreasonably be borne by the public but must be dealt with at source by those who create them. Nor can it be appropriate, it has been argued, for mineral operators, once the material has been extracted, to abandon land, leaving it to other agencies to pay for its rehabilitation.

The development of the framework of controls to deal with these concerns had its origins following the First World War. In the USA federal legislation was passed in 1922 to regulate land use by a system of zoning ordinances. At the time mineral working was outside its scope, but the zoning of other forms of development was a major factor in ensuring that the production of, say, aggregates did not occur in urban or semi-urban environments even though benefits might have accrued to the operator from close proximity to markets. Development controls similar to this spread in the inter-war years to Europe. Even so, it was mainly after 1945 that the closely defined regulation of such matters took hold in such countries as Sweden, Italy, France, Denmark and the UK. In the case of the UK a rather more sophisticated approach was adopted and set out in the Town and Country Planning Act of 1947 (Garner 1975). By the beginning of the 1970s most developed countries had a system of development control which involved the preparation of broad general plans relating to future land use needs usually formulated at the provincial or county level, with much more specifically detailed programmes set out at the level of the municipality, district etc. Moreover, they also had the means by which to consider the desirability of specific development proposals as they were put before them for adjudication. A means of appeal by the unsuccessful applicant for a particular land use change was usually incorporated into the system. Within such broad generalisations regarding the approach to development control there remain wide variations. Although a number of countries apart from the UK including the Scandinavian countries and the German Federal Republic have a comprehensive approach covering all land, it is the exception rather than the rule (Garner 1975).

The system of controlling agencies is also variable and though in most countries minerals come within the purview of ministries of mines, minerals, energy and/or natural resources, in some, departments of environment, planning, construction and even agriculture may be concerned. For a number of nations the responsibility for looking after minerals may vary according to the mineral in question, particularly if strategic considerations are involved. Both Belgium and some Australian states have such an approach.

However, in the case of the UK, although all minerals are handled together there is little centralised control at national or provincial level with responsibility devolved to counties. Only where mineral development proposals are likely to be of a particularly contentious nature (such as the Lee Moor china clay mining application of 1972), are they 'called in' by the Secretary of State for the Environment, who, after receiving the evidence of a public inquiry, will then decide the issue for himself.

The growing environmental debate of the late 1960s and early 1970s in the USA introduced a new element into minerals development which was embodied in the Federal National Environmental Policy Act of 1969. This embraces a dual function of attempting to introduce a means whereby the central issues regarding resource conservation and pollution control may be addressed together whenever a new major official land use development is proposed. These considerations are reviewed in each case through the preparation of an Environmental Impact Statement which sets out the ways in which the proposed development will affect the environment (Catlow and Thirwall 1976). In practice it usually also incorporates a cost benefit analysis of that development. This has offered the opportunity to supplement existing legislation, whether its aim is land use control or the maintenance of public health, and to provide a means of analysing all the factors for and against the proposal. It is an approach which has been copied by many US states, by many Canadian provinces, by Australia at national and state level, and by the German Federal Republic, Denmark, France and Ireland and is under consideration by others (including Norway and the EEC).

Although a wide range of considerations are addressed by Environmental Impact Statements, it should not be supposed that land use planning controls, by contrast, perform that role alone. In the UK context the submission of an application to extract a mineral will cause the recipient local authority's planning department to address considerations which cover its interaction with other forms of land use, and therefore, human activity. If the potential extractor wishes to break new ground not already designated for mineral working the proposal will be referred to other government departments such as the ministries of agriculture and transport, statutory water undertakings and to agencies such as the Nature Conservancy, the Countryside Commission and to the Alkali Inspectorate. Their views as to the desirability of the proposed operation will be made available as will any conditions they feel should be applied to the working if it is permitted to go ahead. For example, the river authority under the River (Prevention of Pollution) Acts, 1951 and 1961, might demand that the disposal of mine or quarry waters should not alter the mineral content of streams or rivers, whilst the Alkali and Clean Air Inspectorate may wish to control dust or other emissions from secondary processing activities under the Clean Air Act of 1956. Agencies controlling mineral development, in whatever ways they exercise their role in the developed industrial countries, are increasingly seeking to apply conditions to consents which demand a degree of site rehabilitation either as a mining operation progresses or upon its completion. Whilst such considerations were first invoked in the USA just before the Second World War in connection with the strip mining of coal in West Virginia, they have chiefly been embraced by countries such as the German Federal Republic, France, some of the states/provinces of the USA, Australia and Canada, as well as the UK in more recent years, although in the case of the last mentioned country the obligation to restore both open cast coal mining and iron-stone sites dates back to 1951. In some instances the government authority has insisted on the deposition in advance of a cash bond which may later be drawn upon for rehabilitation purposes. British Columbia, for example, operates such a scheme and its virtue plainly lies in securing the means of ensuring rehabilitation which is not at the public expense should the extractor become bankrupt or otherwise cease operations prematurely (Blunden *et al.* 1973).

## 8.6  Positive planning for mineral

Although so far this consideration of the controls exercised over mineral working appears from the prospective operators' point of view to be largely negative, it would be a mistake to believe that there is no evidence of the relevant control agencies acting in a positive role. Just as it may be important to reduce dust and noise from a quarry in the interests of other land users, such agencies can ensure that land around a quarry is not taken up with inappropriate uses which could be a source of friction. Land-use planners will wish to avoid locating housing estates adjacent to quarrying operations. Their other important positive role is to ensure that mineral reserves are not sterilised by urban development of the kind that was all too apparent in the inter-war years in the UK. At that time millions of hectares of river terrace gravels were covered by the indiscriminate development of housing estates to the west of London. In the post Second World War situation local authorities have been given much more positive advice from central government regarding their role insofar as it is exercised in relation to the extractive industries:

> In considering whether or to what extent mineral works should be permitted, it is important to bear in mind that the mineral industries are fundamental to the national economy and that many of the other industries of the country are in greater or less dependent on them. The fundamental concern of planning policy must therefore be to ensure a free flow of mineral products at economic cost (Control of Mineral Working, HMSO, 1960).

As a result, certain local authorities in the UK have taken very seriously their obligations towards minerals extraction. In Devon, Cornwall and Dorset, the areas containing workable deposits of china and ball clays are not only allocated for that purpose in their overall structure plans, but adjoining zones are designated as 'consultation areas'. A similar area surrounds the ironstone workings in the Lindsey district of Lincolnshire. This is because such areas are deemed to be of great importance to the working of these minerals. Therefore, a consultation procedure takes place between the industries in question and the planning authority whenever an application for development is received within these areas that does not bear directly on the workings of the minerals concerned. Should the extractors object to the development, the local authority will not let it go ahead without referring it to the Secretary of State for the Environment who will undoubtedly 'call in' the application for decision by himself.

Other authorities have initiated complex exercises in phased planning for particular minerals, based on the best available assessment of future demand, the estimated size of workable deposits and the optimum location of workings to supply the demand. The formulation of special plans such as that of Staffordshire for sand and gravel, or of Somerset County Council for quarrying in the Mendip hills has been welcomed by leaders in the extractive industries. However, such positive approaches have also been made at a higher level. Inventories of valuable mineral reserves have been produced to better co-ordinate their exploitation on a national or provincial basis in most eastern European countries, in Denmark and in some of the German Federal Republic states (Thomas 1980). In other countries such as the UK, Belgium and the Netherlands as well as in some of the Canadian provinces only individual minerals have so far been treated in this way. However, the EEC is undertaking a strategic assessment of its mineral resource base. This exercise could prove an invaluable input to minerals planning at a scale hitherto unknown in Europe and make an important contribution to future discussions leading to Lomé style agreements with African, Caribbean and Pacific less developed countries.

## 8.7 Rehabilitation of mined landscapes

Although modern planning controls over mineral working will almost certainly require some scheme of site rehabilitation when extraction of the valuable material has been completed, what can be achieved in this respect will, of course, vary according to what has been extracted, its location and the feasibility of back-filling waste. Whilst it has been known for underground workings, once completed, to receive back waste rock and tailings, this is usually prohibitive in cost terms. Similarly, it is hardly likely that an open-pit copper mine could at the end of its life be refilled with tailings or overburden, any more than china clay workings would be used for the final resting place of quartz. At the other extreme there will be occasions when a quarry is worked extensively and merely involves the removal of a thin layer of valuable material. Under such circumstances stored top soil and overburden can be replaced and the land returned to other usages.

Such methods are generally used only where the mineral occurs at relatively shallow depths (usually less than 45 metres) and in more or less horizontally bedded deposits. Britain's major sources of ironstone are stratified deposits of this type and with stripping ratios ranging from less than 1:1 up to 16:1 are worked in this way. Another example of such an approach to extractive activity may be found at the Kensworth chalk quarry in Bedfordshire, England, where a shallow band of chalk, used in the manufacture of cement, is removed, the lower surface is resoiled behind the working face and returned to agriculture. Sand and gravel workings are also relatively easily rehabilitated and this again involves the restoration of the landscape at a lower level. If such workings are below the water table, as they frequently are where river terrace gravels are exploited, water recreation areas probably constitute the best form of landscape rehabilitation. Occasionally, where ground water supplies are not likely to be contaminated, refuse tipping has been used to restore levels to match those of the surrounding area.

Hard rock quarries are almost invariably excluded from such methods of working and rehabilitation. However, an exception is to be found in the UK at the Dunbar, Scotland, quarry of Associated Portland Cement Limited. There the raw materials for cement manufacture (limestone and shale) occur in successive horizontal deposits, interbedded with waste rock, down to a depth of 38 metres. The overall stripping ratio is 3:1. A highly ordered and fully mechanised system is used to replace the waste in the correct order. The new land surface is some 12 metres lower than the old and restoration to agriculture is effected two years after quarrying. This type of precision quarrying carries a cost penalty; costs per tonne of stone are about twice those of conventional limestone quarries owned by the company. Nevertheless, in this instance the company clearly felt that these additional charges, caused by the decision of the planning authority to impose such a restoration scheme, were commercially acceptable.

Hard rock quarries in general terms do, however, present more difficult problems in terms of restoration because of the small amounts of waste to valuable material, their depth and their frequent remoteness from urban populations. This often means that it is difficult to find suitable fill material or to transport it there at an acceptable price.

Perhaps amongst the best examples of what can be done to rehabilitate the landscape come from the working of open-cast coal in the UK and in the German Federal Republic where the extraction process involves the removal of shallow horizontally bedded material and causes only a temporary environmental impact. Although the original habitat cannot be re-established, landscape restoration can be more or less complete. Topsoil and sub-soil are first stripped and stacked. Rock and shale overburden is removed and the coal revealed for working. After this is completed, overburden is replaced and graded to prearranged contours to conform to the surrounding land. The increased volume of this material caused by its removal in the first instance is usually more than adequate to make good the 'lost' coal seams.

The subsoil is then spread and rooted to avoid compaction and assist drainage before the topsoil is replaced. If the land is to be used for agriculture, as is usual, fencing, hedges, or walls are placed in position, ditches dug and trees planted. A similar approach is adopted in the restoration of open-cast ironstone workings in the UK.

The most intractable problems stem from the larger metalliferous workings of the open-pit kind, because apart from the presence of the excavation itself, the tailings also can present difficulties. Though as was seen with the tailings at Eliot Lake and Sudbury there may be compelling reasons why such areas are stabilised, they consist of fine material which to any form of plant life offers an environment of high metal toxicity, possible acidity and no suitable nutrients. A vast range of chemical and physical treatments has been applied to re-dress the inherent lack of phosphates, potash and nitrogen and to render the metal ions unavailable to plants. A considerable number of different plant species have been tried subsequently. Although there are now around the world many examples of successfully vegetated tailings areas as a result of such experimental work, the basic unsolved problem remains the deterioration of cover once maintenance ceases.

**CHAPTER 9**

# The positive utilisation of mining wastes

## 9.1  Waste utilisation opportunities

What has been said in Chapter 8 raises the question as to what can be done with wastes in any positive sense outside schemes for rolling restoration. It is a question to which an answer is sought under two main headings: (1) opportunities which may exist for the use of waste as a saleable product, and (2) the use of waste for on site amenity works at the point of their production or as fill elsewhere (Blunden 1980).

The opportunities for making use of wastes from mineral extraction are closely related to the quantities of waste which are available. Depending on the mineral being worked, these quantities at any one extraction site tend to be either very large in relation to the valuable product, or very small. In the first category comes coal as the figures for each country in Table 9.1 testify, and in the UK, the working of china clay and (in the past) slate quarrying are also notable contributors. In the USA, however, where total waste production amounts to over 2,000 million tonnes per year (by far and away the biggest producer of wastes both by quantity and variety), the copper mining industry accounts for almost 50 per cent. Other large quantities of waste, taken in a world context, arise from the working of iron ore and taconite, uranium, phosphate, gold, gypsum, lead and zinc. The second category of small producers of waste include hard rock aggregate enterprises and those of sand and gravel, though here, as Table 9.1 indicates, remarkably little is known about the exact quantities.

## 9.2  Waste as a saleable product

Of the large waste producers, the coal mining industry produces significant quantities of unwanted mudstone, sandstone, shales and kaolinites, together with some carbonaceous material. A brief comparison between the coal waste

211

Table 9.1 Production and utilisation of mines and quarry wastes*

| | Production (Mt. per annum) | | Stockpiles (Mt) | Utilisation |
|---|---|---|---|---|
| | Waste rock (inc. overburden) | Tailings | | |
| *Australia* | | | | |
| Lowgrade ilmenite | 0.2 | — | ? | — |
| Lead/zinc ore | — | 0.5 | 4.75 | — |
| Coal | c.60 | | 3,000 | Small quantities for road construction and fill material (burnt spoil). |
| *Belgium* | | | | |
| Coal | 4 | | — | Lightweight aggregate manufacture. |
| *Canada* | | | | |
| Iron ore | 45 | | — | Aggregate fill; roofing granules. |
| Coal | 5 | | ? | — |
| *Finland* | | | | |
| Misc. operations | 3.1 | 5.1 | 94.8 | Road construction; aggregate for concrete. |
| *France* | | | | |
| Coal | 20 | | 700 | Approx. 35% production used for: (i) road construction – fill material; (ii) lightweight aggregate manufacture; (iii) brick making. |
| *Germany* | | | | |
| Coal | 63 | | 100s? | Approx. 35% production used for: (i) road construction – fill and embankments; (ii) land fill. |

| | | | | |
|---|---|---|---|---|
| **Holland** | | | | |
| Coal | 0.3 | | — | 40% production used for block making (from fired mixture of colliery waste, flyash and sawdust). |
| **India** | | | | |
| Laterite | 2 | ? | — | Clay pozzolana: lime substitute in mortar. |
| Gypsum | 0.2 | — | — | Building plaster. |
| China clay | — | — | — | Clay pozzolana; brick making. |
| Mica | | 0.005 | | Mica insulation bricks. |
| Coal | 8–10 | | | Mine backfilling (large %); road construction; fill. |
| **New Zealand** | | | | |
| Coal | ? | ? | ? | Small quantities used for land reclamation and fill. |
| **South Africa** | | | | |
| Gold ore | 4 | c.40 | — | 50% waste rock used for road construction and aggregate (for concrete); silicate bricks (minor usage). |
| Misc. quarrying | | | | Waste rock used for road construction and fill; tailings in silicate bricks. |
| Coal | 9 | | 150 | Very little used. |
| **Sweden** | | | | |
| Iron ore | 25 | 9.6 | 289 | Land fill (limited use); road construction and brick making (small quantities). |
| Sulphite ore | — | 11.7 | 57 | |
| **UK** | | | | |
| China clay | 22 | | 300 | Less than 5% production used in: (i) aggregate for concrete; (ii) silicate bricks; (iii) road construction; (iv) fill; (v) amenity banks. |
| Slate | 1.2 | | 300–500 | Inert fillers and roofing felt (small % of total used). |
| Tin ore | 0.5 | | c. 0.30 | Aggregate for concrete (minor usage). |
| Fluorspar | 0.23 | | — | Road construction and aggregate for concrete (minor usage). |

| | Production (Mt. per annum) | | Stockpiles (Mt) | Utilisation |
| --- | --- | --- | --- | --- |
| | Waste rock (inc. over-burden) | Tailings | | |
| Misc. quarrying | ? | — | — | Road construction; silicate bricks; amenity banks. |
| Coal | 50–60 | | 3,000 | 11–13% production used in: (i) road construction (fill and building); (ii) brick making; (iii) cement manufacture; (iv) lightweight aggregate manufacture. |
| *USA* | | | | |
| Copper | 624 | 234 | 7,700(a) | Road construction; bitumen filler. |
| Taconite | 100 | 109 | 3,600(a) | Aggregate-skid resistant. |
| Phosphate Ore | 230 | 54 | 907(a) | — |
| Iron ore | 27 | 27 | 730(a) | Road construction. |
| Gold ore | 15 | 5 | 450(a) | Road construction; aggregate for concrete. |
| Uranium ore | 15.6 | 5.8 | 110(a) | Aggregate for bituminous concrete. |
| Lead ore | 0.5 | 8 | 180(a) | — |
| Zinc ore | 0.9 | 7.2 | 180(a) | — |
| Misc. quarrying | 68 | — | ? | Amenity banks. |
| Gypsum | 14.2 | 2.7 | ? | — |
| Asbestos | 0.6 | 2 | 14(a) | — |
| Barite | 1.9 | 3.1 | 25(a) | — |
| Fluorspar | 0.1 | 0.4 | ? | — |
| Feldspar | 0.2 | 0.8 | ? | — |
| Coal (bituminous) | 100+ | | 2,000 | Small quantities used in: (i) road construction – base and subbase materials; surfacing aggregates; (ii) cement manufacture; (iii) mineral wool manufacture. |
| (anthracite) | 1 | | 700 | Anthracite waste used in: (i) concrete block making; (ii) brick making; (iii) manufacture of lightweight aggregates. |

(a) Stockpiles for USA are tailings only.

214

of the United States, Germany and Australia shows that although the chief mineral components are somewhat similar, the last named has a significantly higher carbon content. As a result of upgrading, a major use for it is found in the construction of roads. Where it is utilised for the building of embankments especially in German, France and the UK, the material is first burnt. Burning shale waste improves the strength of the material along with its resistance to weathering. This burnt material is also widely used in the UK for road sub-base and base construction, but unburnt spoil can also be effective if mixed with cement, or, as in France, with fly ash, lime or gypsum.

Colliery waste as fill beneath buildings is widely used in the UK where new developments are in close proximity to supplies. Problems can arise if the material is unconsolidated or if the sulphate and acid content of the spoil is allowed to make direct contact with concrete, a problem which can easily be avoided. Furthermore, where colliery waste or shale occurring naturally contain large quantities of pyrite or calcite, the material may swell if it is allowed to oxidise, a situation which has been identified in Canada, the USA and the UK. Advice on how to avoid such problems when using shale is given in the Building Digests of Canada and the UK.

Colliery spoil is also used in brick-making in France, the UK and to a lesser extent in the USA, whilst extensive trials in its use for this purpose are in progress in Germany. In France production reached 200,000 bricks a day by 1983. In the UK, however, competition from concrete blocks and from the highly cost effective brick clays of the Vale of Oxford which result in a product of better finish but equal durability, have largely banished the spoil bricks to Scotland, remote from the clay brick fields of Bedfordshire, Buckinghamshire and the Peterborough area.

The manufacture of synthetic aggregates from colliery spoil occurs in France, the UK, Belgium and Poland and probably other Eastern European countries. In the USA waste from anthracite mining is used in the manufacture of concrete blocks. The sinterstrand process (a low-cost and low-energy method of synthetic aggregate production) used in the UK and Poland produces a rough angular material well suited for use in light-weight concrete blocks but not for structural concrete. Those employing a rotary kiln technique as in France and Belgium produce a material suitable for both concrete blocks and for structural concrete manufacture.

Finally, colliery spoil is also used to replace the clay fraction in the feed to cement works. It supplies the alumina, silica and iron necessary for the formation of cementitious minerals as well as contributing to the fuel re-

---

* Adapted from proceedings of an international symposium on waste material sponsored by the Réunion Internationale des Laboratoires d'Essais et de Récherchés sur les Materiaux et les Constructions (RILEM), published in *Materials and Structures*, **12**, No. 70, 1979.

*Source*: Gutt, Nixon (1980) and other sources.

quirements of the process. Its utilisation in this context is, however, small and appears to be confined to the USA and the UK.

Apart from these uses, research into other positive outlets for colliery waste is proceeding. In Australia, Germany and the UK the fluidised combustion of colliery spoil, particularly tailings, is being developed both for the production of heat energy and ash. The latter is useful as a high-grade fill; as an aggregate for road making; in light-weight aggregates used to produce concrete blocks; as a grog in the making of bricks, tiles and ceramics; as a soil conditioner; as a material for filter beds; and for the absorption of oil slicks and spillages.

In Germany, not content with using colliery spoil in the lower layers of roads and embankments, research is being carried out to upgrade it to a level at which it can be used in the upper layers and for the frost blanket. In the UK promising experiments are in hand to use colliery spoil alone to produce skid resistant road-surfacing material as well as mixtures of this spoil and bauxite.

In all cases, except where colliery waste is upgraded via a relatively sophisticated process and considerable value added to it, its utilisation in some positive way is largely constrained by the comparatively short distances over which it may be economically transported.

Of the other forms of waste produced in substantial quantities, these mainly derive from metalliferous mines or as a result of extracting other industrial minerals. The type of material available for positive utilisation will consist either of waste rock that is excavated to expose the valuable mineral during mine development (overburden), or the residues obtained from the separation of valuable minerals from their ores (tailings). In all instances, their chemical composition and size will vary widely. In the USA, although the size of waste rock from mining is generally less than 0.3 metres, residues from the processing of ores where fine grinding is practised will be in the claysilt range with 50–90 per cent of such tailings under 75 $\mu$. By comparison, waste from the winning and processing of china clay consists of overburden, coarse sand with a particle size of 9 mm to 75 $\mu$ and a fine micaceous waste of particle size 75 $\mu$ to 10 $\mu$ in the rough proportions 4:4:1.

## 9.3 Upgrading wastes

In terms of upgrading, relatively little use is made of the materials emanating from mining operations other than those of coal, partly because of their natural limitations. The possibility of the toxic contamination of water leaking through tailings containing heavy metals is a severe constraint on their use even as fill. For both phosphate and uranium wastes, where they have been similarly used as preparation for housing development, abnormally high levels of radioactivity have been observed. Evidence on this latter point has been obtained in Canada, the USA and Australia (Gutt, Nixon 1980).

Asbestos waste, another hazardous product in terms of its possible impact on human health, has, fortunately, not been used as a fill or in any other context.

On the other hand, evidence from the UK indicates that the coarser wastes from china clay workings as well as from fluorspar have been used in roadmaking in all layers, from skid-resistant surfacing to fill material as well as in concrete making where they act as an aggregate. In Australia manganese mud (the waste from the production of manganese by electrolysis) is used to colour bricks, whilst in India wastes from china clay are used to make bricks or are burned to form pozzolanas.

Of the finer materials, these are used as fillers in bitumen and in the manufacture of autoclaved silicate products. Mine tailings are being experimented with in Australia and the USA to determine their suitability for the making of ceramic products, and in the latter country suitable investigations have taken place into the utilisation of such wastes in calcium silicate products, bricks, light-weight building blocks and mineral wool. Because of the considerable availability of slate wastes in the UK, these have been used in a ground form as a filler in plastics and paints and for roofing granules. Work is in hand to use slate waste as an aggregate in the making of concrete bricks. Also the manufacture of autoclaved aerated materials from slate powder and Portland cement has been investigated. Using a mixture of 45 per cent slate powder with cement plus aluminium powder as an aerating agent, aerated concrete has been produced with properties similar to that currently available by standard production methods.

The processing of phosphate also produces considerable quantities of fine material (75% under 3 $\mu$) in the form of slimes. In Florida, where the USA industry is mainly located, between 9 and 13 million tonnes per annum of these slimes are produced, with over 2,000 million tonnes already stored in ponds. There is considerable pressure for their utilisation in some positive manner. Fortunately, phosphate processing also produces a silica sand. Although this can be readily disposed of as fill or used in concrete, when mixed with the slimes the result is a stable material which can be used as a land fill. The ratio of slimes to sand prevents more than a third being used in this way. Alternative utilisation for the slimes has been in the production of light-weight aggregates but the market needs developing. Unfortunately where the stabilised slimes have been used as fill under houses in Florida, radioactivity has posed a problem.

Quarrying industries are at that end of the scale where waste production is very small in relation to output. In India, wastes from laterite workings are burnt to form pozzolanas or to make bricks. In the UK, major companies such as Imperial Chemical Industries and Amey Roadstone Corporation have looked at the use of the clay fraction of limestone quarry wastes as a raw material for cement manufacture. The latter has also examined the production of foundry and filtration sands from quarry waste. But in these cases the problem is one of *insufficient* waste materials, since only the very largest quarries are ever likely to be able to support, for example, a cement kiln. Although it is interesting to speculate about the possibility that several adjacent quarries might erect a joint kiln to dispose of their clay wastes, there seems little likelihood of this solution being implemented.

The changing conditions of supply and demand are having some effect upon the utilisation of scalpings (dirty stone rejected after primary crushing) at quarries. Previously considered as waste, or saleable only at minimal price within the immediate locality, an increasing shortage of local aggregate sources in, for example, the south-east of England, is resulting in the long distance haulage of scalpings for sale as bulk fill. Indeed, scalpings are now railed from Somerset to Southampton (96 kilometres) and to London (160 kilometres). In some cases, old scalpings piles are being re-processed to obtain clean stone. A factor of some importance in the UK is that mineral operators pay local authority taxes on sales. Thus, the sale of waste materials – such as china clay sands – increases the companies tax outgoings and may act as a positive disincentive to the pursuance of alternative uses. This effect would be reversed by a tax on waste disposal instead of waste sales, but there is no immediate prospect of such a change.

Returning to a world-wide perspective, the uses to which wastes can be put do not result at present in significant inroads into their current production, let alone the vast backlog from earlier output. Consequently the bulk of these wastes is tipped in the traditional manner. One reason for this is that often the wastes in question are located in places well away from the major settlement areas and transport costs can make them uneconomic in relation to competing materials. This, together with the inherently small size of the market for upgraded wastes, gives little reason to believe that such alternative uses for these wastes will in the short or medium term solve the problems of their disposal to possible economic and environmental advantage.

## 9.4 Waste for amenity works

It is now extremely common for quarrying companies in EEC countries and North America to build amenity banks to screen all or part of their operations. In a UK sample survey of sixty major operations, nearly 70 per cent had constructed amenity banks of differing scope and size (Blunden 1980). Such banks are normally constructed from overburden and waste rock. They differ from waste tips insofar as they perform a screening function and, ideally are not themselves obtrusive. Normally they are built to provide a visual screen by intercepting particular sight lines. Occasionally such banks may also act as noise baffles but they have seldom been built specifically for this purpose and would need to be located close to either the noise source or the recipient to fulfil this function. Banks may also have some value in lessening dust emission from the site.

Because the banks are constructed from waste materials, they have the same disadvantage as waste tips, that is the appropriation of land. However, in the case of banks built on a sufficiently large scale, proper planning can ensure that productive land use is re-established when the bank is completed.

Numerous factors require attention during the design and construction of an amenity bank. In most cases professional advice is needed to ensure that the end result achieves the objective. The location, form and contouring of the bank must achieve the desired screening effect. Unless the form of the bank is in harmony with adjacent land forms, the bank may have a visual impact as great as the feature which it conceals. The land form should also be designed to suit the intended after use. If, for example agricultural machinery is to be used on the bank, excessive slopes need to be avoided. Such slopes will, in any case, hinder vegetation establishment. Finally the requirements of its surface and its total stability need to be met.

A further important consideration is that, during the life of the extractive operation, it is mainly the external face of the bank which determines the success of the structure. However, at closure and rehabilitation of the site, the appearance of the internal face of the bank may also be important. It will be most effective if both faces are designed to a satisfactory standard, rather than attempting to remedy the inner face at a later date.

The use of appropriate waste materials and their careful placement in the bank is essential if stability and vegetation problems are to be avoided. Waste most suited for vegetation needs to be reserved for the final cover of the banks. Conversely, hostile or toxic waste may be buried within it, thus effecting their satisfactory disposal. The likelihood of water pollution due to leakage must receive attention.

The establishment of vegetation on the bank involves several considerations. First, there is the choice of species. If the bank is to be incorporated into an existing agricultural holding, choice of plant species will be conditioned by the requirements of the farmer. If the bank is to fulfil a visual requirement then species appropriate to the local natural habitats may be used. Second, to avoid a situation in which the bank has an undesirable impact of its own, it may be advisable to replicate or augment existing vegetation forms. For example by extending natural woodland with tree planting on the bank, the bank may be successfully incorporated as a natural landscape feature. Finally, planting needs to be carried out at the correct season by competent specialists.

For any given site particular constraints will exist that result in the completed bank being a compromise between the requirements of land form, availability of waste materials used for building the bank, and vegetation. Whatever the exact features of the bank finally constructed, maintenance will be required to ensure its long-term success. Any slumping or erosion of the soil surfaces has to be promptly remedied. Vegetation needs to be cut, weeded or fertilised as appropriate. Any failed vegetation needs to be replaced if erosion is to be avoided.

219

The positive utilisation of mining wastes

The construction of amenity screens to hide extractive activities at or near their site will, therefore, often provide an environmentally acceptable way of using these relatively small quantities of waste in a productive manner. Operators are divided on the question of the significance of the cost involved in such schemes when compared with the total cost and profitability of their business. Examples of the order of magnitude of costs incurred are given in Table 9.2 and relate to a study carried out in the UK, by the Mining Environmental Research Unit (Blunden et al. 1974).

These costs require discussion. In the case of quarries A and B, the expenditure figures given are solely related to the cost of grading and/or establishing vegetation. No costings for earth moving to form the basic bank structure are included. In comparison with the annual operating surpluses, the amounts spent are minimal – still more so when related to the estimated working lives of the quarries concerned. The costs incurred by quarry C are however, of a wholly different order of magnitude – 33 per cent of estimated gross income for the year in which the bank was built. This bank covers just under three hectares and contains over 500,000 tonnes of overburden. It was built by contractors within six months and both these factors increased the cost. The $460,000 includes all the costs associated with the bank from drilling, blasting and loading the material, to transport, tipping, grading and revegetation. Apart from grading, revegetation and a part of the transport costs, all these operations would have been needed to form the equivalent waste tip that working would have demanded. The major part of the expenditure would thus have been incurred anyway in waste tipping, so that the true cost of the amenity bank is very much less than it appears; in this case no more than about one-fifth. Even this amount may be inflated because of the use of the contractors to build the bank. On a per hectare cost-

Table 9.2 Costs of amenity banks

| Quarry | Output (tonnes per year) | Estimated life (year) | Estimated gross income ($ per year) | Estimated operating surplus ($ per year) | Costs of amenity bank ($) |
|---|---|---|---|---|---|
| A | 700,000 | 70 | 2,300,000 | 322,000 | 13,570 |
| B | 1,000,000 | 15 | 2,645,000 | 460,000 | 10,860 |
| C | 500,00 | 60 | 1,380,000 | 253,000 | 460,000 |

Notes:
Quarry A: Costs of amenity bank (0.4047 hectares) were $11,270 for grading a pre-existing tip and $2,300 for seeding and stabilisation.
Quarry B: Costs of amenity bank (0.607 hectares) are solely revegetation costs. Earth moving and re-grading costs are not included.
Quarry C: The high cost of this bank conceals a number of specific factors discussed in the text.
Source: Blunden et al. (1974).

220

ing, therefore, the bank was probably little more expensive than at quarries A and B, and as in those cases, was only a small proportion of the estimated operating surplus. Regardless of the significance which is attached to these costs at any particular site, it is common ground among operators that such schemes involve expenditure on which no cash return is obtained. On the other hand, the wider community has received benefit from the works undertaken. Some operators realise that, although unquantifiable by known techniques, the community and land use planning authority good-will which may be generated is likely to result in some commercial advantage. This assumes, of course, that the works were not insisted upon as an integral part of the original consent to operate the quarry. Although research is being devoted to such topics as the financial value of landscapes, it is unfortunately insufficiently far advanced to be utilised for a quantification of the value of an amenity scheme at a mineral working.

## 9.5 The local transport of waste for use as fill

Those situations where mining wastes may be used in the immediate reclamation of those sites where they were produced was addressed in Chapter 8. But there it was acknowledged that some extractive activities will produce excavations which cannot be filled because wastes to valuable product ratios do not permit. The converse state of affairs has also been recognised. In this section the extent to which surplus mining wastes at one site may be used in the reclamation of others will therefore be considered.

Abandoned excavations at or within a short radius of working properties (15–25 kms) may be advantageously filled with wastes from current production. This course of action eliminates the need to take virgin land for tipping and, if the original topography is more or less restored, will result in a useful and valuable new area of land. It is possible that over short distances the cost of transport will be offset by the avoidance of land purchase for tipping and the value of the reclaimed land.

Prior to undertaking such work, the nature of the abandoned excavation requires examination. If it is flooded or if fissures are in continuity with subterranean water, water pollution may be caused if the wastes contain silts or clays. The type of waste to be used also requires consideration. If, at completion of refilling, the area is to be revegetated, it will be desirable that wastes most conducive to vegetation establishment be reserved as top soil. Top soil from overburden stripping is best kept for this purpose.

Hitherto the disposal of wastes into excavations has been a haphazard process depending upon the free choice of the waste producers. In the UK, however, it has become increasingly felt that this *ad hoc* approach is no longer sufficient. Accordingly, the Northern Group of the Local Government Operational Research Unit (1973) carried out a study in the 'Five Towns' area of West Yorkshire (Leeds, Wakefield, Dewsbury, Castleford

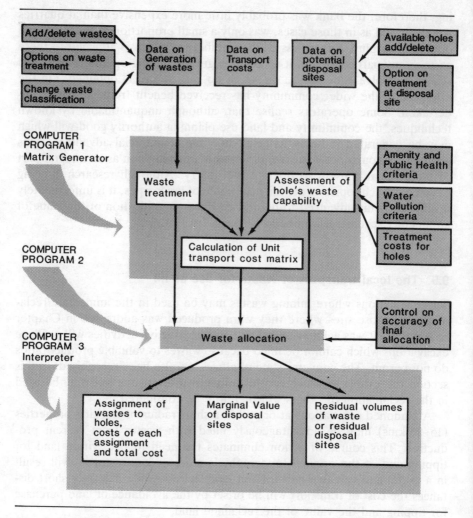

*Fig. 9.1* Computerised allocation model for the least-cost association of wastes and disposal sites. *Source*: Local Government Operational Research Unit (1973)

and Pontefract). This study aimed at locating and typifying all waste-production sites in the area, waste-production rates, and actual and potential disposal sites. Mathematical techniques in conjunction with a model were used to allocate waste to the most economical disposal sites, taking into account the suitability of different wastes for different sites, quantities, transport methods and costs and so on. A computer program for use in routine decision making on this topic has been produced (Fig. 9.1). No insuperable difficulty is seen to exist in extending this on a national basis or applying it in areas of other countries, particularly where there is a high level of urbanization.

However, this is not to imply the absence of other studies concerning rehabilitation, most notably that of McLellan, Yundt and Dorfman (1979). In this the dimensions of the problem for Ontario, the most highly populated of Canada's provinces, have been postulated in the number, location and characteristics of all abandoned pits and quarries. Here alternative reclamation uses for each of these are established whether they be forestry, recreation, agriculture, housing or commerce, the final goal being a strategy whereby land owners and community planners are in a position to propose suitable rehabilitation programmes to be paid for from a proposed provincial rehabilitation fund where commercial considerations cannot be expected to apply. This study, although not as sophisticated in some senses as that of the Local Government Operational Research Unit, is clearly more pragmatic in its approach, leaving an element of negotiation in the final decision-making process regarding rehabilitation, but at the same time identifying priorities in terms of the need to rehabilitate certain sites in advance of others. A strong methodological element concerning how to approach the problem of rehabilitation rationally is again in evidence as it was in the West Yorkshire study.

## 9.6 Long-distance transport of waste

There remains the possibility of removing waste materials from the extraction point to some more remote location for use in a major project. If for the immediate future reprocessing and upgrading is ruled out, the only obvious use in this context would be in a large-scale public works programme of land reclamation. To the author's knowledge, no such major scheme has yet been undertaken though feasibility studies are in hand in the UK for the transport of colliery waste from adjacent coal fields to reclaim sites on the Firth of Forth in Scotland and Humberside in England. Although data is not available concerning these schemes, a fully detailed study of the large-scale application of the long-distance transport of waste materials arose in connection with the proposal to locate London's third international airport at a coastal site at Maplin in Essex (Blunden 1980). The land needed 400 million tonnes of fill if it were to be reclaimed. Possible sources were the china clay wastes of Cornwall and Devon and the slate wastes of North Wales, but the logistics and costs of transporting these materials in excess of 400 kms suggested that colliery waste might be preferred. This was suggested in a report produced by the National Coal Board in the early 1970s when the Maplin area first came under scrutiny as a site for a major airport. This envisaged the movement of the 400 million tonnes of waste over a ten-year period. Some 200 million tonnes would come from the Yorkshire/Nottinghamshire coal fields (about 220 kms) and 100 million tonnes from both the north-east (over 360 kms) and south Wales (about 320 kms). Further material would be available if required. At these sources, a total of over 46 sq. kms of serious dereliction would be removed and the land obtained would be capable of re-use. Overall, the delivery costs of waste would have averaged $3.5 per tonne at 1973 prices.

The total cost of using colliery waste for the reclamation work was thus $1,318 million. This compared very unfavourably with the estimated $320 million to use locally dredged materials. Taking a working figure of $5,683 per hectare as the value of re-instated colliery land, the net cost of using coal wastes was around $1,035 million. This sum represents the effective cost of using Maplin to remove 46 sq. kms of derelict land, equivalent to an approximate cost of over $227,000 per hectare. When it is realised that some quite expensive landscaping and revegetation work on colliery waste tips *in situ* was carried out at the time of the National Coal Board's report at over $17,000 per hectare, it is extremely doubtful that the vast cost of removing the waste would result in benefits that could not have been obtained for a more modest expenditure.

Another example of long distance transport of wastes but again with reference to the UK, comes from the china clay industry. Corner and Stafford (1971) prepared for the Devon County Council a report on the disposal and long-distance transport of china clay wastes from the Lee Moor workings. This study demonstrates several important features of the problem. Total stockpiles of sand reached 28 million tonnes in the year 1979/80 with local uses of the waste only disposing of about 8 per cent of each year's production. Meanwhile experiments in the manufacture of artificial aggregate from the sand showed in the early 1970s a minimum price of $4.60 per tonne, or about four times the price of natural aggregate prevailing at the time. It was concluded that none of the existing uses of the waste would absorb more than a minute proportion of the total.

A study was therefore undertaken to test the possibility of selling the material in south-east England as an alternative to local sources of natural minerals. The only feasible modes of transport were rail and sea. Direct road haulage costs were not competitive with rail for distances in excess of 120 kms. To take two examples: to transport to Southampton, a distance of over 240 kms by rail would have cost $4.07 per tonne by that mode but $5.06 by sea; to Gravesend east of London, a rail distance of around 400 kms, would cost $5.60 by either mode. Overall, the price differential in favour of locally excavated materials would have been about $3.20 per tonne, and there were in any case doubts that the south-east could absorb the quantities of sand available without closures of many existing pits.

Although not of direct relevance to the minerals industry, the UK brick clay fields of Buckinghamshire, Bedfordshire and around Peterborough are another good example of the problems encountered in the long-distance transport of waste for reclamation purposes (Blunden 1975). In the Peterborough area, more than 400 hectares of pit are available for filling. A tripartite reclamation scheme was begun in 1966. The Central Electricity Generating Board, after spending $8.5 million on terminal and other facilities, commenced railing power station fly ash to fill pits owned by the London Brick Company. To restore the back-filled pits to agriculture, top soil was supplied by the British Sugar Corporation which obtained it from the washing of sugar beet. The cost of railing ash from between 80 to 95 kms to Peterborough was about $1.20 per tonne. However a scheme to rail colliery waste over a 140 km distance to these clay pits has not been implemented due partly to transport costs of $1.70 per tonne. Costs of this order of magnitude, totalling some $5.5 million per annum, would have been unacceptable if borne solely by the four collieries from which the waste was to have come. Investigations into alternative transport modes suggested that the use of pipelines would reduce the figure to no more than $0.28 per tonne at then current prices; nonetheless no action had been taken as of 1983.

Even the most ambitious public works programme could, at best, remove only the more serious areas of dereliction due to waste disposal. If, within the context of the UK, it were decided to use all backlog wastes from the coal, china clay and slate industries in such a manner, the equivalent of nine Maplin airports would have to be built, together with another airport every five to six years to cope with current production.

## 9.7 Policies for waste utilisation

When considered in conventional economic terms, the long-distance transport of large tonnages of waste is seldom likely to be attractive. Hence it is unlikely that mineral operators will ever voluntarily undertake such movements. Equally, there are few alternative uses of upgraded wastes which are economically viable in competition with natural products, particularly if significant transport costs are incurred to reach the markets. Conventional economics does not take account of unquantifiable benefits such as visual improvement, the lessening of nuisance or danger to the public, and so on. It is thus when these factors appear important that there is scope for government intervention on the basis of a policy that could decide whether the benefits of waste removal are worth the apparent 'loss' in money in doing so. It seems therefore that in the case of small-scale waste production (as at quarries) and in view of the fact that many opportunities for environmental improvement work at extraction sites still exist, current and future waste production will be largely devoted to on-site amenity work. For major schemes of waste transport and utilisation, government support would be required.

However, the formulaton by individual countries of any policy for the positive utilisation of mineral waste is hampered by the lack of quantification of existing waste stocks, and current and future production, as well as insufficient information on the properties of the wastes and their suitability for different purposes. Nor does any national picture exist within those countries of the relative locations of waste heaps and the excavations to which they might be returned. The task of making such studies in a country such as the UK is feasible because of its small size and important because of its shortage of land arising from a population density of 227 persons per square kilometre as compared with 22 persons in the United States of America. Even so, within North America and elsewhere there are areas that may be considered to be heavily urbanised where such studies could be equally valid. Until they are undertaken so that the relevant data may be obtained, proper planning for waste utilisation will not be possible.

# CHAPTER 10

# Minerals exploitation and regional development

## 10.1  Alternative resource development opportunities

The definition of a resource implies its having intrinsic utility or value to man. Thus the exploitation of any resource must inevitably lead to increased employment opportunities and to economic growth for the areas in which it is exploited and even to other areas well beyond.

In highly developed and diverse economic regions the development of a hitherto unexploited resource may have only a marginal impact. The first of the two case studies in this chapter illustrates this. However, that which differentiates mineral resource development from that pertaining to other resources is its transient nature. This situation may be demonstrated by reference once again to this same first case study of a deposit of tungsten found as a disseminated low-grade ore associated with kaolinised granite.

Assuming an annual output of up to 4,500 tonnes of ore, the tungsten resource will have an exploitable life of twenty years, with site rehabilitation occurring shortly after the beginning of the twenty-first century. Yet the mine site is on the south-western boundary of Dartmoor, an area which has long been valued for its scenery and the recreational facilities it provides. Since 1951 the area has been designated as a National Park and enjoys the landscape protection and management afforded it by the National Parks Access to the Countryside Act, 1949. As the network of motorways has extended westwards and such an area has come within a three hour journey time from major conurbations of Birmingham and London and opportunities for leisure have markedly increased with rising living standards, then its importance and value in such a role has increased and so far as can be ascertained this is likely to remain the situation.

The gradual increase in the appreciation of Dartmoor National Park as a recreational resource has at the same time offered enhanced opportunities for employment in the area, far greater than those likely to accrue to the working of a single resource such as tungsten, or the other major mineral extracted in the area, kaolin. These opportunities accumulate directly to those who service the needs of the visitors (hoteliers, guest-house keepers, shopkeepers), as well as providing additional employment for those who are responsible at a local level for the management of the park. The rapid circulation of money derived from recreational activity in a rural area such as Dartmoor may be contrasted with that earned by the exploitation of tungsten since the percentage of the total expenditure that is incurred by the latter but which goes out of the area is likely to be very much greater. Thus, whilst the local impact of a mining activity should not be understated and will indeed be explored in greater detail below, compared with those of recreation and tourism which are based on a continuous resource, its 'life' in terms of development opportunities is constrained. In the case of the tungsten mine it is within a generation and though other stock resources may have a longer duration, most point sources of these are capable of total exploitation well within the framework of normal human experience. Conversely, Dartmoor, as a recreational resource, though it may be affected by man's actions, like other such resources, it can be there in perpetuity and certainly very long after the winning of a mineral such as tungsten has ceased.

## 10.2  Mining and regional infrastructures

The impact of mining on regional development during the lifetime of the working will vary greatly according to the nature of the mineral resource in question, the amount of investment involved, the mining techniques deployed and the level of employment generated by it. Variations will also occur over time – as mining development involves different factor inputs according to the stage in its development – and over space; remote locations for example require a heavy investment in infrastructure to provide water, power and transport, together with those items required for the social care of the workers and their families such as education, housing and health and community services. Extreme cases occur where such infrastructural developments exceed the cost of the mine at the construction stage. The Liberian Lanico iron ore mining project with an ore output of 7.5 million tonnes per year entailed an outlay of $100 million. Half this was spent on harbour and railway construction; and 10 per cent on electricity supply as well as the building of a township and a hospital. Only 32 per cent of the total sum was spent on prospecting, planning and the construction of the mine (McDivitt and Manners 1974).

In other instances, not only may infrastructural developments be entirely absent but the regional impact of the mine may be minimal or its effects may leak away. Baldwin (1966) has described the failure of Zambian metal mining to generate any real regional impact since its inception in the 1920s. The mining has been highly capital intensive, no local industries have been set up to utilise the output of the mine, nor have they provided any significant numbers of the indigenous population with incomes large enough to transform consumption patterns for local goods and services or to create food production in the area. Spooner (1981) has recognised the failure of the Appalachian region of the USA to benefit from the exploitation of its mineral wealth and other natural resources such as timber. The Appalachian Regional Commission has drawn attention to a lack of reinvestment of the wealth obtained from an exploitive economy as a key factor in the poverty of the area. Bosson and Varon (1977) have noted the propensity of some mining developments in the less developed countries to remain enclaves with comparatively little effect on their surrounding regions, though their contention that only 30 per cent of minerals won are processed in such countries, 'a proportion that has remained constant since 1950', was by the 1980s no longer proving to be the case as Chapters 5, 6 and 7 on the production on individual minerals make clear.

In less developed countries forward linkages are increasingly becoming apparent within such regions. For metallic ores beneficiation processes involving concentration, pelletisation or sintering are now common although smelting is still rather less so. As we have noted such developments greatly increase the value of the product to the producer nation and to the region. With iron ore alone pelletisation or sintering can more than double the value per tonne of ore produced (McDivitt and Manners 1974). Another but less common forward linkage in less developed countries involves the localised acquisition of mineral-using industries, a most important element in the process of regional development.

Other impacts of mineral development in less developed countries involve what may best be described as 'back-wash' effects. Friedmann (1973) has reported that the Chilean nitrate and copper development enclaves in the northern region of the country have a powerful impact on the national economy. Tax revenues from the mining corporations operating there have been spent with less than wholly desirable results in the capital of Santiago. Hyper-urbanisation has been stimulated, diverting scarce resources away from more meaningful investment schemes in industrial and agricultural enterprise with real benefits accruing mainly to the banking and commercial classes. In Venezuela, according to Odell (1973), the Maracaibo oilfields could be seen as isolated high-technology enclaves in a region of subsistence agriculture throughout the 1940s and 1950s. The oil was piped away with little or no local impact since the international oil companies undertaking the exploitation of the oil were totally self-sufficient. However, taxation revenues were expended on the development of the capital, Caracas, where the oil companies also had their headquarters. By the 1960s, apart from the misspending of such revenues on non-essential building in that city, such policies had resulted in a severe imbalance of the population within the country with some 25 per cent of the Venezuelan people clustered in the Caracas metropolitan area.

229

## 10.3 The post mineral development phase

Whilst there may be a varied response in development terms resulting from mining activities during their production stage, there may equally be considerable differences in the response the regions may make to the demise or substantial contraction of mining operations. Warren (1973) cited a number of instances of this situation taken from North America and the UK. As interesting as his examples of the ghost towns of the Wild West of the USA is his reference to Camborne and Redruth in the UK. These two Cornish towns, as he pointed out, grew to serve the needs of the metal mining activities of the area (mainly tin and copper). They attained their greatest importance around 1871 when tin production reached its peak (11,100 tonnes of metal). Although in the 1970s the tin mining industry began to reassert itself as a fact of economic importance in the county (its output in 1980 was worth about £22 millions), production is now only about one-third of what it was a century ago in a greatly expanded market. Moreover, its productive capacity is currently based on a small number of capital intensive workings compared in the past with a multiplicity of small labour-intensive enterprises. Consequently, Camborne and Redruth are towns largely depleted of the functions they were originally designed to serve with populations too great for the present resource base. They are now primarily associated with tourism.

Blakemore (1971) cited the example of northern Chile where the agricultural exploitation of non-irrigated areas was historically associated with the mining camps. When the veins of metal were worked out, the agricultural villages declined. However, elsewhere examples may be found of the tendency of mining developments to stimulate local supportive activities which may last long after the disappearance of the original stimulus to regional economic expansion. In California the goldrush of 1849 led to the influx of labour not only to work the mines, but also to serve the needs of those who worked them by opening up large crop and livestock areas which eventually became independent of the mining industry (Warren 1973). Even though the Cornish metal mining industries declined and remained depressed throughout the inter-war years, many ancillary enterprises set up to supply mining equipment, explosives and safety fuses have been maintained and now supply world-wide markets. This is a palliative in an area of above-average unemployment.

More significant and spectacular examples of the propensity of mining to sustain growth exists in the case of many of the British coal fields. Brown (1972) has argued that coal 'had an original and dominating role as a source of energy' and was also 'a big enough activity in itself for its direct distribution to have direct economic significance'. The coalfields were able to draw in a huge labour force and at a time when transport costs were comparatively greater, exercised a powerful pull on other industries. It has been estimated by Hall (1973) that in 1900 over 50 per cent of British towns were situated on or near coalfields. He also stressed that each coalfield ultimately sustained 'a host of varied crafts and trades' and that their pull was only reduced by the development in the mid-1930s of a national electricity power grid.

Although the association between coal as a resource and its power to stimulate other industrial activities has been largely broken, the desire to utilise the social and economic fabric resulting from that situation has been recognised in the influx of non-coal related industry, though frequently aided and abetted by government grants and other local incentives. The propensity of mining activity to do more than merely create direct employment at the exploration, the development and the exploitation stages, is, however, worthy of more careful analysis than that afforded by the descriptive stance so far taken in this chapter. To take a more quantitative approach to two specific but contrasted case studies, the use of multiplier models is deployed.

## 10.4 The development of theoretical and employment multiplier models

Although multiplier models involve many complex problems in their practical application, the basic concept is simple and deterministic. Let us suppose, for instance, that there is an initial injection of incomes into an area via a mining project requiring several hundred men (mainly local manual workers) and perhaps twenty to thirty other persons for clerical and other indoor jobs. At least some of the money generated by this enterprise would be spent locally and would find its way into the pockets of local businessmen who, in their turn, would spend some of their extra income locally with other businessmen, and so on. The total amount of local income generated will be greater when more is spent locally and less is allowed to leak away through the purchase of goods and services from outside the area. Thus it is possible to estimate in any situation the hypothetical size of the local multiplier effect, given assumptions about local spending and after making allowance for marginal rates of taxation and marginal rates of savings.

The impact of a new venture also may be measured via its effect upon total employment. Again the initial increase in employment would generate needs for additional work in the locality and these extra workers would in turn increase the demand for other local employment and so on. Once again it is possible to project a theoretical final outcome to this multiplicative process given certain assumed relationships between inputs and outputs for the various sections of the local economy.

A relatively simple multiplier mechanism of this type pertaining to mining activity is illustrated in Figure 10.1. Four types of relationships are important: (1) purchases of inputs; (2) purchases by the labour force employed; (3) induced investment in the infrastructure which may come about, not so

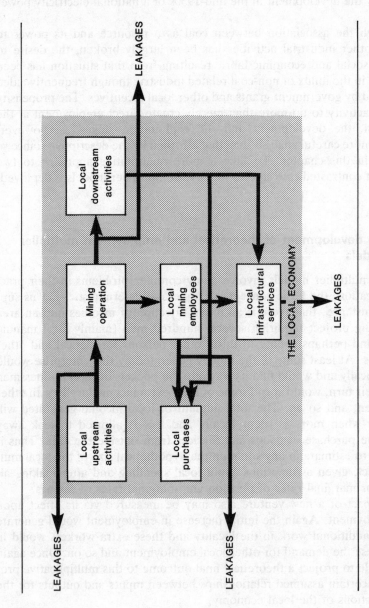

Fig. 10.1 A simple multiplier model. Source: Blunden (1982)

much by the employee purchases as by demands for extra central government services; and (4) 'downstream' linkages which can involve a further build-up of employment.

Taking 'upstream' or input activities first, a mining operation might use local services such as transport, maintenance, plant hire, and construction and civil engineering workers and consultants. Other services which might be bought locally, but more likely would be 'imported' from outside the defined locality, would include explosives, chemicals, mining plant and equipment. The greater the amount to be bought locally, the greater the boost to the local economy since less is allowed to 'leak' away on imports.

Another form of input concerns administrative and higher level functions – purchasing, marketing, research and development, operational research, personnel and industrial relations and so on. These services might be internalised within the local operation, bought locally, or – more likely – be provided by a head office of a multi-national mining corporation spatially divorced from the local mining operation. Again, the more of these functions carried out locally, the better for the indigenous economy, particularly because of the range and quality of work involved.

In the short run, employee spending will have an impact on the local community in a way which depends upon a number of factors. The net effect of spending by newcomers to the work force, either local school-leavers or previously economically inactive females, or immigrant workers, is likely to be greater than the effect of employing a local married man already receiving substantial levels of unemployment and other benefits. A substantial influx of new workers and their families may generate demands for extra private and public-sector facilities. However, imported workers may bring with them different spending patterns which do not benefit local businesses. An extreme example is the immigrant who sends virtually his entire earnings back to his area of origin. But it is also possible for increased earnings to bring about a change in the spending habits of local workers. With more money in their pocket they may go further afield to spend it.

In the long run, when all the effects of a new operation have worked their way through an economy, demands of a socio-political nature may arise for improved public-sector services – schools, hospitals, social services and so on. These may give rise to an increase in the amount of public sector residentiary employment in the area.

In general, however, multiplier effects are more likely to be greater, the greater the proportion of immigrant workers. But this is not, of course, necessarily a desirable feature, since employment of local people has a job-preserving function. An injection of work which preserves employment will thus also support infrastructural services and residentiary employment and prevent a downward spiral of labour demand. Turning to 'downstream' activities, a mining enterprise, as was seen in Chapters 5, 6 and 7, can and increasingly does engender considerable local processing activities and smelting; indeed manufacturing enterprises using the finished product may also arise.

To summarise this brief review of theoretical considerations, the impact of a mining operation might bring about considerable upstream and downstream activities. The combined effect of all these will increase the demand for further residentiary services, and in time may result in an inflow of exogenous funds from central government or private sources to finance further infrastructural facilities. The spin-off will depend upon the ability of local inhabitants to provide upstream and downstream services, and the spatial spending patterns of employees and employers.

## 10.5 Case study 1: Tungsten mining as an income and employment multiplier

Against the background of the foregoing explanation of multiplier models, the real world and the two case studies can be examined. The first concerns the proposal to develop by the mid-1980s an open-pit tungsten mine at Hemerdon on the south-western edge of Dartmoor National Park in Devon, England. The area in question abuts that shown in Fig. 1.1 of Chapter 1 and is therefore close by the extant china clay workings of Lee Moor. The tungsten ore to be worked is found at an average 0.143 per cent disseminated in a matrix of china clay with tin (averaging 0.026%) and waste quartz and micaceous residues. The proposal is that once the development and construction phases of the mine reach completion, the open pit will be worked at an extraction rate of up to 4,500 tonnes of tungsten concentrate per year during the life of the mine. The purpose of applying the model to such an example was to make a contribution to the evidence put forward at the public inquiry (September/October 1982) regarding the environmental acceptability of such an enterprise by commenting on the value of the mine to the area in terms of the creation of additional wealth and jobs (over and above those directly employed at the mine) in its surrounding area. Apart from the collection of data about wages and salaries, numbers employed at the mine, their place of domicile as well as that pertaining to all the other non-labour outgoings of the mining operations, the prime consideration in carrying through the work was the definition of the study area or sub-region. Other similar studies of sub-regions have been concerned with fairly small and well-defined watertight communities. Examples include the study of the multiplier impact of a projected copper mine on Anglesey, an island off the coast of North Wales (Blunden 1977), and those for the Scottish universities at Stirling and St Andrews (Brownrigg 1972; Blake and McDowell, 1967). Unfortunately, the setting for the present study does not meet those criteria. The bulk of the labour will be drawn (see Table 10.1) from the adjacent Plymouth area with a population of around 300,000 which is projected to rise over the period of mine operations to 350,000 with a workforce of 125,000 rising to 150,000. Consumer purchasing is dominated by the Ply-

*Table 10.1*   Hemerdon Mining Project – workforce

| Forecast origin of operational workforce | | | |
| --- | --- | --- | --- |
| Plymouth | Plympton | Sparkwell | Other nearby villages |
| 40% | 30% | 15% | 15% |

*Source*: Blunden and Perry (1982).

mouth built-up area with around 2,000 shops, whereas neighbouring local shopping centres – such as Ivybridge – only possess some 35–50 shops.

Any attempts to exclude Plymouth/Plympton from the study area would result in an enormous leakage of expenditure out of the area chosen and give rise to minimal local multiplier values. In other words the purely local impact in employment or income terms of the mining operations would likely be very small. On the other hand, the leakages from a wider area including Plymouth/Plympton would likely be quite small, with high multiplier values. Having said this, it has to be acknowledged that the actual contribution of a relatively small project to a fairly large urban area might hardly be felt, useful though it may be.

The study area was, therefore, defined as a zone including the built-up area of Plymouth and extending some kilometres to the north, west and east of the mining operation. This part of the areal boundary was ultimately not delineated with accuracy though the overwhelming focus of activities is towards Plymouth, while other areas are unlikely to be much affected. The study area thus defined is very similar, in terms of population and workforce, to that used in a series of detailed studies of the multiplier impact of industrial activity in four districts of Cornwall undertaken by Perry (1982). This made it appropriate to use those findings as a yardstick for assessing the results of the work from the defined Plymouth study area.

### 10.5.1 Multiplier estimates – the construction phase

The construction phase of any venture differs from that of the operational stage in many important ways; the quantity and quality of labour used, the type of plant and machinery involved, the demands upon the infrastructure and, perhaps most important of all, the ephemeral nature of the impact. Some construction projects can leave an itinerant workforce behind them, but this should not happen at Hemerdon since the bulk of the labour is envisaged as being recruited locally (Table 10.2) and the number of immigrant semi- or unskilled workers, who might pose the greatest unemployment problem if they remained in the area, will be very limited (Table 10.3).

As to the employment of outside contractors, both small builders and civil engineers are to be used at the construction stage. In the Cornish studies already cited, the local multiplier effect of small building firms was found to give one of the highest values of any industrial sector. No evidence was found within the Plymouth area that contradicts these findings. One of the

*Table 10.2*  Hemerdon Mining Project: Construction Phase – labour

*Use of local and outside labour*

|  | Year One (1st half) | Year One (2nd half) | Year Two (1st half) | Year Two (2nd half) |
|---|---|---|---|---|
| Local | 65 | 115 | 195 | 175 |
| Outside | 15 | 50 | 135 | 125 |
| Total | 80 | 165 | 330 | 330 |

*Source*: Blunden and Perry (1982).

main reasons for their high multiplicative performance derives from their relatively low levels of tax 'leakage' compared with their larger counterparts and their extensive use of local commercial, professional and other services, notwithstanding their relatively labour-intensive nature compared with other sectors. By contrast larger civil engineering companies with their headquarters outside the districts covered by the Cornish studies did not use local commercial and professional services although they did make use of local vehicle and plant repair facilities, a situation closely paralleled in the Plymouth sub-region.

As to the construction at Hemerdon itself, in addition to normal construction and civil engineering work, a further 70 people, half-skilled and half-unskilled, are likely to be employed locally upon small-scale fabrication works during the second half of the two-year building phase. Thus, at the peak of the construction stage around 400 people may be engaged upon the project, comprising 60 managerial and clerical staff and 185 skilled and 155 semi- and unskilled workers. Of the total, 135 seem likely to be imported workers. As in most projects of this type the proportion of staff recruited from outside increases with skill content – about 10 per cent of the unskilled, 25 per cent semi-skilled, nearly 40 per cent of the skilled and 50 per cent managerial and technical.

The total payroll of the construction phase is scheduled to rise from £1.1 million in the first year to £3.75 million in the second. All costs are quoted at constant 1981 values. The proportion of total costs accounted for by non-labour components is in line with the Cornish survey data and is only very briefly described here (Table 10.4). Company vehicles are for local use. Diesel is for site use and of course contains a very high 'import' content in

*Table 10.3*  Hemerdon Mining Project: Construction Phase – imported labour

*Imported semi or unskilled workers*

| Year One (1st half) | Year One (2nd half) | Year Two (1st half) | Year Two (2nd half) |
|---|---|---|---|
| 10 | 20 | 35 | 15 |

*Source*: Blunden and Perry (1982).

*Table 10.4*  Hemerdon Mining Project: Construction Phase – non-labour inputs (in £ sterling)

| | Year one (1st half) | Year one (2nd half) | Year two (1st half) | Year two (2nd half) |
|---|---|---|---|---|
| Electricity | 20,000 | 20,000 | 30,000 | 50,000 |
| Water Authority | 1,000 | 1,000 | 1,000 | 1,000 |
| Rates | 5,000 | 5,000 | 15,000 | 15,000 |
| Post/Telecoms | 10,000 | 10,000 | 10,000 | 10,000 |
| Transport | 5,000 | 5,000 | 5,000 | 5,000 |
| Company vehicles | 25,000 | 25,000 | 25,000 | 25,000 |
| Diesel (for on-site vehicles) | 60,000 | 60,000 | 60,000 | 100,000 |
| Insurance | 15,000 | 20,000 | 40,000 | 50,000 |
| Equipment and machinery | — | 200,000 | 300,000 | 300,000 |
| Consultancy | 75,000 | 75,000 | 75,000 | 75,000 |
| Hotels/catering | 20,000 | 20,000 | 20,000 | 20,000 |
| Accommodation | 66,200 | 35,000 | 112,500 | 112,500 |
| Totals | £302,200 | £476,000 | £693,500 | £763,500 |

*Note*: 1981 prices.

*Source*: Blunden and Perry (1982).

its cost. Insurances are placed externally. Equipment and machinery also has a very high 'import' content but about 50 per cent of consultant's fees are expected to be earned locally. The 'hotels/restaurants' figures exclude accommodation for non-local on-site personnel which is estimated separately. No local housebuilding – or infrastructural improvement – effects are included in the calculations because it is assumed that the relatively short stay of immigrant workers will be absorbed into existing accommodation (about 30 staff in hotels and 70 to 80 workers in lodgings).

It will be evident from what has been said about non-labour inputs to the project in its construction phase that the total expenditure on these cannot be included in the calculation of the total injection of income into the area. This is because of 'leakages' due to the 'importation' of some goods and services and for this reason reductions have to be made. In the case of the first year non-labour costs, after an appraisal of each of the relevant factors, the 'leakage' is considered to be of the order of £528,000 and the total non-labour costs of £778,000 given in Table 10.6 is reduced accordingly. For the second year the 'leakage' is considered to be about £707,000 and the total non-labour costs figure of £1,457,000 is required to be reduced by that amount. Thus the direct injection of income in the construction phase was estimated at £1.35 million in the first year (being the 'adjusted' non-labour cost plus the payroll cost) and £4.5 million in the second.

Assuming that the bulk of the work is actually carried out by smaller local sub-contractors, even though it is handled by larger outside contractors, the overall income multiplier is estimated at 1.4, in line with that of the previously cited Cornish studies. This value is similar to that calculated for the projects in rural Scotland mentioned earlier and somewhat higher than estimated for mine construction work in Anglesey, a figure of 1.3 (Blunden 1977). This higher prediction was felt to be justified since the degree of 'leakage' through consumer spending is likely to be less in the Devon venture. The 1.4 multiplier value yields a forecast of total injections of income into the local economy of £1.9 million in the first year (being the 'adjusted' non-labour cost plus the payroll cost × the multiplier 1.4) and £6.3 million in the second year. The employment multiplier impact was, however, assumed to be minimal, since it is contended that the local infrastructure could absorb the diffused extra load in terms of demand for accommodation and private and public-sector facilities, utilities and services.

### 10.5.2 Multiplier estimates – the operational phase

Since the production life-cycle of the mine is projected to cover a period of twenty years and since the initial run-up as well as the final run-down period is expected to be very short, the estimates made here were intended to be typical of the whole life-cycle of the operation. (All cost figures are at 1981 prices.) It is perhaps worth comparing the life-span of the mine with that of new manufacturing starts in the area. During the 1961–73 period, the death-rate of Devon and Cornwall enterprises was at about the UK level of 2–3 per cent per annum which might suggest a life expectancy of 30–50 years. However, the failure rate rose sharply in the later 1970s and the surveys have suggested that new firms in the study area are now over-dependent on a narrow product range and lack regenerative ability. It was expected that this attribution can be made to Plymouth-based firms, although one study found evidence of a slowness to adapt to changing markets, technology and methods (Hankinson 1978). It is possible therefore that the average expectation of life of a new manufacturing enterprise may be similar to that for the mining project.

Table 10.5 gives the projected skill structure of the labour force for the venture. A comparison with the data taken from the surveys of Cornish industry by Perry (1982) indicates that the skill composition of the mining company is slightly better than the norm for Cornish firms. However from the evidence available, it seems likely that Plymouth units generally employ higher proportions of managerial and skilled staff.

The annual payroll is estimated at £2.8 millions. As Table 10.5 illustrates, nearly a third of the workforce is projected as recruited from outside the area. They would earn about a quarter of total wages and salaries. From evidence collected in the area, local consumer spending is assumed to be at a high level, rather greater than that estimated for the Cornish surveys and even higher than the coefficients calculated for the remoter parts of Wales

Case study 1: Tungsten mining as an income and employment multiplier

*Table 10.5* Hemerdon Mining Project Operational Phase – staffing

| | Number | Average salaries/wages (£) | Recruited Locally | Outside |
|---|---|---|---|---|
| Clerical and office | 14 | 5,000 | 14 | — |
| Management/super-visory | 32 | 12,150 | 21 | 11 |
| Technical specialists | 26 | 8,500 | 18 | 8 |
| Skilled operatives | 83 | 8,900 | 59 | 24 |
| Semi-skilled operatives | 123 | 8,200 | 95 | 28 |
| Unskilled operatives | 62 | 6,100 | 51 | 11 |
| Total | 340 | — | 258 | 82 |

*Note*: As is made clear on p 231, in calculating a local multiplier effect which is in part based on salaries and wages earned at the mine, allowance must be made for taxation, other deductions at source, and savings which do not support the local community infrastructure.

*Source*: Blunden and Perry (1982).

(that is the Anglesey study) and those of Grieg covering the impact of a pulp and paper mill in the Scottish Highlands, 1971, and Scottish coastal fishing, 1972. Non-labour costs are summarised in Table 10.6 from which it can be seen that they total £5.4 million p.a. or about two-thirds of the total costs, in line with the findings of the Cornish survey (Perry 1982). In Table 10.6 'fuel' comprises all operation needs. 'Insurance' is placed externally. 'Hotels/catering', refers to business visits to the area while 'business travel' comprises visits elsewhere, mostly by air. About a half of all consultants' fees are expected to be earned locally.

In the Cornish surveys of multiplier impacts, it seems that the mining sector performed rather better than the rest of the industrial base since 'imported' items from outside amounted to only 30 per cent of total inputs for metalliferous mining, compared with between 45 and 55 per cent for different manufacturing units. Thus although metalliferous mining in the Cornish areas seems to have a better multiplier impact than the manufacturing industry as a whole, it would appear that the results for tungsten mining in the Plymouth sub-region would be even better because of an assumed lower leakage of consumer spending out of the area and somewhat better skill and salary structure mentioned above.

Thus an income multiplier value of 1.7 was posited compared with the Cornish estimate of 1.6, for mining, Grieg's (1971) Scottish figure of 1.5 and also the Anglesey figure of 1.5. It follows that where the payroll costs were concerned, a multiplier value of 1.7 would transform the annual total figure of £2.8 million p.a. into a total local impact of £4.77 million p.a.

*Table 10.6*  Hemerdon Mining Project: Operational Phase – costs of services, rates, other services and supplies

| | |
|---|---|
| 1. *Service and rates* | |
| (a) Gas | Nil |
| (b) Electricity | £1 million p.a. |
| (c) Water Authority | £10,000 p.a. |
| (d) Rates | £170,000 p.a. |
| | |
| 2. *Other services and supplies* | |
| (a) Post and Telecoms | £25,000 p.a. |
| (b) Transport | Business travel £10,000 p.a. (80% air, 20% road). There would also be an estimated 20 heavy goods vehicles visits per day by non-Amax owned vehicles of which say 50% would originate locally: cost £25,000 p.a. |
| (c) Company-owned transport | 10 cars + 10 light commercial vehicles £50,000 p.a. |
| (d) Insurance | £120,000 p.a. placed through London brokers. |
| (e) Equipment and machinery | Mine plant £900,000 p.a. |
| | Steel balls/rods for grinding £130,000 p.a. |
| | Chemicals £150,000 p.a. |
| | Maintenance supplies £200,000 p.a. |
| | Packaging £90,000 p.a. |
| | Plant hire £75,000 p.a. |
| | Fuel £1,250,000 p.a. |
| | Explosives £900,000 p.a. |
| | Miscellaneous £130,000 p.a. |

Of this, say 75% of maintenance supplies and 100% of plant hire and fuel would come from local suppliers.

| | |
|---|---|
| (f) Professional and Scientific Consultancy | £110,000 p.a. of which 50% local. |
| (g) Building maintenance | £50,000 p.a. all local. |
| (h) Hotels/catering | £10,000 p.a. all local. |

*Source*: Blunden and Perry (1982).

As to non-labour expenditure dealt with in Table 10.6, if all the services and goods, and so on, mentioned here could be obtained within the Plymouth sub-region then the impact of the multiplier would yield £9.18 million and when added to the payroll impact of £4.77 million p.a. it would give an estimate for the total income effect upon the area of £13.95 million annually. However, in the case of non-labour expenditure items, it can be noted that they may be divided into outside supplies and services (such as electricity and fuel) on the one hand, and site plant equipment, maintenance and professional and other localised services on the other. Table 10.6 shows that the great bulk of the cost of outside supplies and services is accounted for by electricity and fuel and the 'leakage' of revenue from this kind of expenditure outside the Plymouth area is likely to be very great. As far as hardware was concerned, the largest expenditure was allocated to equipment which will be brought in from outside the Plymouth area. Again, only a small percentage margin for local dealers/suppliers/contractors is likely to accrue locally. Thus, after examination of each non-labour cost as shown in Table 10.6 in terms of likely levels of 'leakage', a total of £581,000 p.a. was allowed to remain in local hands out of the total £5.4 million annually, or 10.76 per cent of total non-labour expenditure a year. Therefore the total estimated local income impact of the mine during its operational phase will be the total payroll effect of £4.77 million p.a. (that is the actual payroll figure of £2.8 million × the multiplier 1.7) plus the non-payroll effect of £987,700 (that is the 'adjusted' non-payroll figure of £581,000 × the multiplier, 1.7). This amounts to about £5.76 million annually.

No reference has been made to the downstream linkages of tungsten working simply because there are none in relation to the sub-region. In this respect the mining of minerals may well be less favourable to regional development than other forms of development, a point underlined by the spin-off effects recognised by Grieg (1971 and 1972). For the Scottish pulp and paper mill development, the initial creation of 850 jobs led to the further employment of between 300 and 500 people. With the Scottish fishing developments, every 100 new off-shore fishing jobs created between 138 and 330 on-shore jobs in fish processing, packing and marketing.

Given the assumption of depressed manufacturing in the Plymouth sub-region, it was forecast that indirect employment effects would not be so great, and also that the incoming 82 workers (of whom 19 will be managerial and technical workers and probably diffused over a wider area) would be absorbed into the local infrastructure without generating needs for infrastructural development. The longer-term multiplier effect of the operation is best seen in terms of employment preservation in both the primary and tertiary sectors. This is because the most likely scenario for the Plymouth sub-region over the next twenty years is that manufacturing employment will not increase at a sufficiently fast rate to offset local unemployment plus a labour supply swollen by the type of inward migration of working-age population encountered in the south-west over the past two decades. Even if the area continues to enjoy special government financial support by virtue of its Assisted Area status, it is very probable that the relative emphasis upon the periphery will be much weakened in favour of the central regions with their mounting problems of inner-city decay, pollution and violence. On the assumption that the proportion of the working population employed in the residentiary sector will continue to rise, the most likely outcome is that the 340 people employed in the mining enterprise will maintain a demand for as many workers again in the residentiary services. In total this will mean employment provided for some 650 people in the area, rising to almost 700 through the life of the project.

This conclusion is based on the evidence of the ratio between industrial

241

and residentiary employment calculated by Perry (1982) for Cornwall for the last twenty years. Although at the beginning of the period the ratio was 100:99, by 1971 this had fallen to 100:90. This change occurred at a time when basic employment (manufacturing and tourism) was expanding fairly quickly (by 70% a year) while residentiary employment appeared to be falling (by 3%). Later in the first half of the 1970s basic employment continued to increase but with residentiary employment now moving ahead even faster. By the mid 1970s basic employment was falling but with residentiary employment continuing to grow, much more in line with UK trends, although well short of USA ratios (100:210 according to Garnick 1970). Thus, the prognostications postulated for added residentiary employment in the Plymouth sub-region at 1:1 appear to be extremely conservative.

## 10.6 Case study 2: Oil as an employment and income multiplier

The development of Scottish oil, as with other stock mineral resources which are traded internationally and are of high value, has implications of significance at two levels, national and regional. The difference is one of scale and perspective. At the national level, the question of balance of payments and taxation revenue is undoubtedly the most immediate and key issue. At the regional level, employment opportunities and the importance of the mineral in terms of regional income are paramount both now and in the future.

In examining the direct impact of Scottish North Sea oil exploitation, Mackay and Mackay (1975) have distinguished four overlapping phases: (1) exploration; (2) manufacturing (that is the fabrication of production facilities); (3) construction; and (4) production. Their 'construction' phase parallels the others and includes such temporary activities as pipe laying. For each field (in 1983 there were fourteen viable fields from Montrose to the Shetlands) the second phase of manufacturing plus some construction is the most labour-intensive and has lasted between three and seven years. By the time the production phase was reached, one which may last some thirty years, employment levels had fallen from their highest point with the industry now best described as capital intensive. But as production rose in the early 1980s to reach maximum levels of between 100 and 150 million tonnes of crude oil per year and the nation became a net exporter with the five interrelated fields in the area east of Shetland meeting all British needs, the servicing of the on-site platforms and the on-shore installations has become of major significance, a factor not recognised by Mackay and Mackay.

Reverting to their first stage, however, this was largely contained in the period from December 1964, when the first allocation of licences for Scottish waters was taken up, to 1976. It was necessarily marked by the emphasis given to drilling and to the proving of identified fields followed thereafter by preparations to bring oil ashore. During part of this time a specific pattern of oil-related employment began to emerge with two major concentrations discernable, one around Aberdeen embracing a wide range of activities and the other in the Cromarty–Moray region largely devoted to platform construction as the industry became more involved with oil production than with field development. The most striking feature of this emerging pattern is that most of the oil-related activity has been and will certainly remain in areas far from the traditional main centres of Scottish industry and population. It is for this reason that the development of this new resource is particularly far-reaching in both economic and social terms. Indeed, its effects have been increasingly felt in areas primarily associated until recently with fishing, farming and crofting.

To reach the production levels of the early 1980s for the Scottish oilfields it was estimated that between £3,000 and £4,000 million had to be invested (Blunden 1977). Some of this cash was needed to service the capital raised on the international money market, and a further amount was required to purchase specialised equipment mainly from the USA. However, it is also thought that about half the total development sum was expended in Scotland on the coastline adjacent to the off-shore developments; that is, in those areas along the east coast already identified, especially Aberdeen which effectively established itself as the 'off-shore capital of Europe'. A considerable part of this money was spent on construction platforms, though a range of other equipment has been needed including around a hundred supply vessels costing £1 to £1.5 million each and designed to service the North Sea rigs and platforms. About ten pipelaying/derrick barges have also been required, along with four survey vessels and eight berthing/fire-fighting tugs. Although these were by no means all built in Scotland, or even the UK, the subsequent maintenance of these has brought benefits to this north-east coast. It is the service infrastructure of the area which has more recently been undergoing expansion and that will be most important in terms of the on-going exploitation of the oil resource. The nature of these services as they are required throughout the development of the oil fields is shown in Figure 10.2. However, the location of service bases is governed by accessibility to the oil fields and the developing pattern can in this respect best be understood by reference to the range of service vessels and helicopters and the economics of supply operations. Since just over 300 kms seems the upper limit for the effective operation of service vessels, as well as the limit of the range of helicopters, bases have been required to be opened up not only at Aberdeen to serve the well heads from the Forties field down to Argyll, but more recently in the Shetlands (at Sullom Voe) to serve the more northerly fields of Thistle, Dunlin and Brent (see Fig. 10.3). The development of the latter is, however, unlikely to detract from the pre-eminence of Aberdeen which, because of its infrastructure as the leading regional city and its close proximity to the earlier discoveries (that is the Forties fields), gained a head-start in terms of oil-based development. Apart from the fact that BP and Shell have their North Sea exploration headquarters in Aberdeen, the number of supply vessels using the port facilities rose from 259 in 1969 to over 2,000 by the mid-1970s. Amoco, Shell, Texaco and Total all exclusively op-

243

| OPERATIONS arranged in order of sequence of development (1 to 6) | accommodation and catering | | | ports | | transport | | | | | |
|---|---|---|---|---|---|---|---|---|---|---|---|
| SERVICES AND FACILITIES — individual categories | housing | work camps and other temporary accommodation | contract catering, laundry services, etc | all weather harbours | very deep water close inshore | service vessels and tugs | cargo vessels and barges | rail transport | heavy road haulage | scheduled and charter air services | helicopter services |
| 1 geophysical and oceanographic exploration | ○ | ○ | ○ | ● | ○ | ○ | ○ | ○ | ○ | ● | ○ |
| 2 construction of service bases and ports | ○ | ● | ● | ○ | ○ | ○ | ● | ○ | ● | ● | ○ |
| 3 oil rig operation | ● | ○ | ● | ● | ○ | ● | ○ | ● | ● | ● | ● |
| 4 pipe laying — sea | ○ | ○ | ● | ● | ○ | ● | ● | ○ | ● | ● | ● |
| 4 pipe laying — land | ○ | ● | ● | ○ | ○ | ○ | ○ | ○ | ● | ○ | ○ |
| 5 production platform construction | ● | ● | ● | ○ | ○ | ○ | ● | ● | ● | ● | ○ |
| 6 production platform operation | ● | ○ | ● | ● | ○ | ● | ○ | ● | ● | ● | ● |
| management of exploration and production | ● | ○ | ○ | ○ | ○ | ○ | ○ | ● | ○ | ● | ● |

○ not required  ● required  N.B. Other existing local services which have to be greatly expanded to serve the oil industry include: hotels, licensed premises, restaurants, car hire, taxi services, banking and insurance, secretarial services, travel agencies, employment agencies, property agencies (lawyers and estate agents). Building and construction trades are particularly heavily employed with labour shortages quickly developing.

*Fig. 10.2*   Services and facilities required by various oil/gas exploration and production activities. *Source*: Hutcheson, Hogg (1975)

erate their support facilities from that harbour and a regular cargo-liner link has been established to the US Gulf ports carrying equipment to and fro. In terms of the redevelopment of the harbour a £2 million investment programme has been completed to provide additional docking facilities and to make the port facility useable at all states of the tide, and new warehousing and offices for supply vessels account for a further £10 millions of capital provided from public and private funds.

| communications | | | supply handling and storage | | | | engineering/ engineering supplies and services | | | | | | | | |
|---|---|---|---|---|---|---|---|---|---|---|---|---|---|---|---|
| stand by safety vessels | navigational aids | telecommunications | open storage | warehousing | specialized storage (silos, tanks, etc) | freight handling, customs documentation, etc | dredging | land reclamation | machinery repairs and servicing | steel fabrication | plant hire (cranes, compressors, welding & drilling equipment, etc) | industrial gases | muds and mineral fluids | drilling consultancy services | diving services |
| ○ | ● | ● | ○ | ● | ○ | ● | ○ | ○ | ● | ○ | ○ | ● | ○ | ● | ○ |
| ○ | ○ | ○ | ● | ○ | ○ | ● | ● | ● | ● | ● | ● | ● | ○ | ○ | ● |
| ● | ● | ● | ● | ● | ● | ● | ○ | ○ | ● | ● | ● | ● | ● | ● | ● |
| ○ | ● | ● | ● | ● | ● | ● | ● | ○ | ● | ○ | ○ | ● | ○ | ○ | ● |
| ○ | ○ | ● | ● | ○ | ○ | ● | ○ | ○ | ● | ○ | ● | ● | ○ | ○ | ○ |
| ○ | ○ | ○ | ● | ○ | ● | ○ | ○ | ○ | ● | ● | ● | ● | ○ | ○ | ● |
| ● | ● | ● | ● | ● | ● | ● | ○ | ○ | ● | ● | ● | ● | ● | ● | ● |
| ○ | ○ | ● | ○ | ○ | ○ | ○ | ○ | ○ | ○ | ○ | ○ | ○ | ○ | ● | ○ |

With an industry which is not only spatially dispersed but also enjoys a number of identifiable development phases some of which overlap, estimates of the precise growth in employment opportunities in north-east Scotland are difficult to quantify. In the specific area of Aberdeen which has longest enjoyed the impact of oil and oil-related growth through all of its development stages, it is apparent that prior to its impact on the local economy (for example, 1966) unemployment rates ran well above the national average. By October 1971 this had fallen to 3.6 per cent, in October 1973 it was down to 1.6 per cent and stayed near that figure throughout the 1970s. In the northeast of Scotland in general wage rates were also well below the national average prior to North Sea developments. Although this has largely remained true in agriculture and food processing, by the early 1970s they had become above average in the building, construction and secretarial service sectors.

245

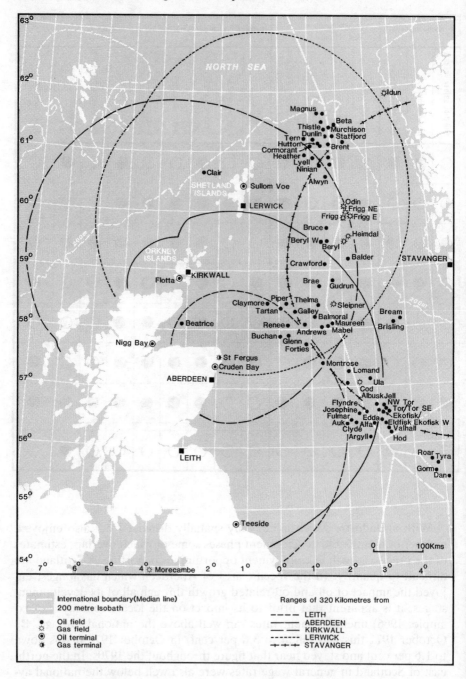

*Fig. 10.3* Accessibility to the North Sea oil fields from key port locations. *Source*: Blunden (1977)

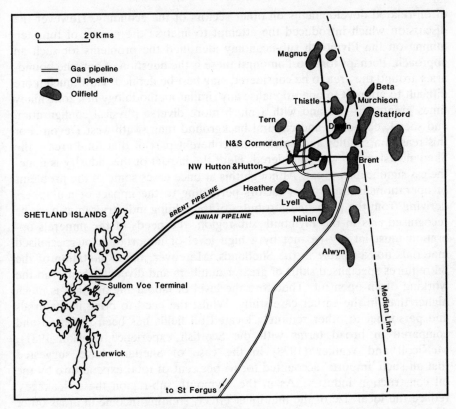

*Fig. 10.4* North Sea oil fields associated with the Sullom Voe terminal. *Source*: Petroleum Review (1981)

The 1971 census figures showed for the first time in this century a net flow of population to this part of Scotland. The north-west of Scotland Development Authority figures indicate that companies were moving into the area during the last decade at an average rate of one a week. Over 200 oil-related companies are now based in Aberdeen and employment resulting from off-shore business had risen from negligible levels to 6,500 in the years up to January 1975 and continued to grow through the rest of the 1970s at around 1,500 per year.

### 10.6.1  Multiplier estimates – the construction phase

So far the discussion of the impact of the oil industry on the economy of the north-east of Scotland has been couched in broad terms identifying for a considerable area its effects at the construction, production and the subsequent servicing stages. Specific quantification of its impact other than through direct employment requires an assessment of the multiplier effects

247

of oil-related developments on other sectors of the economy. However the discussion which introduced the attempt to analyse the impact of tungsten mining on the Plymouth sub-economy identified the problems for such an approach. Perhaps foremost amongst these is the question of how the boundaries around the area to be considered may best be defined. How much more difficult to attempt to operationalise any similar methodology in a zone many times greater in size and with a much more diverse physical configuration and social, economic and cultural background than south-west Devon. For this reason a smaller and much more coherent part of that total area – the Shetland Islands – was considered. Here the impact of the industry is nonetheless significantly great. Although this at once eases some of the problems of operationalisation, other factors pertaining to the impact of multipliers deriving from the particular attributes of the mining industry can at once be recognised. As in the Plymouth sub-region, the needs of the minerals operation must largely be met by a high level of importation of specialised materials not available in the Shetlands. Moreover, the exploitation of the oil requires specialised skills of greater numbers and diversity than even the working of an open pit. Therefore the level of labour importation is much higher than in the earlier case study. Whilst the need to bring in materials and personnel to other remotely located oil fields has been reviewed and compared in broad terms with the Scottish experience (Nelson 1981), McNicoll and Walker (1978) in the case of Shetland have suggested that all such 'imports' accounted for 66 per cent of total expenditure by the oil construction industry. As in the Plymouth sub-region these 'leakages' reduce the local size of the multiplier effect, creating income in areas other than the Shetlands. Moreover, it also must be remembered that at the construction stage of the Sullom Voe terminal when the importation of workers was at its height, these 'temporary' residents spent much of their income outside the Shetlands either by exporting a large part of their earnings to their homes or by expenditure during their leave periods.

Nevertheless the considerable impact of oil-related development on local industry can be demonstrated. McNicoll and Walker (1978) have shown that for 1976 construction work for the oil industry generated a total output of £32.8 million in Shetland, which approximated to over a third of the total output of all local-based industry. The industrial sectors most dependent on oil construction were distribution, transport and that group of 'other services' which covers hotels, restaurants and motor vehicle servicing (Table 10.7). In each of these the additional output generated was in excess of £1 million. Though reaping benefits from oil construction of about half of those groups quoted above, the quarrying industry was over twice as dependent on such construction activities for its output. In terms of employment generation three quarters of the jobs which were deemed to have been created by oil construction were in transport, distribution, communications and 'other services'. To determine the extent of these oil construction-created jobs it was assumed by McNicoll (1980) that a proportionate relationship

*Table 10.7   Output and employment generated by Shetland oil construction in 1976*

| Industry | Oil construction-generated output (£000) | Oil construction-generated output as % of total industry output | Oil construction-generated employment |
|---|---|---|---|
| Agriculture | 92.3 | 4.1 | 25 |
| Fishing | 15.9 | 0.4 | 2 |
| Quarrying | 545.7 | 60.9 | 25 |
| Fish processing | 14.2 | 0.2 | 1 |
| Textiles | 5.3 | 0.2 | 1 |
| Ship repair | 41.6 | 4.5 | 4 |
| Other manufacturing | 319.4 | 17.9 | 22 |
| Construction | 289.0 | 5.1 | 37 |
| Utilities | 275.5 | 16.6 | 13 |
| Transport | 1,618.2 | 27.0 | 167 |
| Distribution | 1.937.1 | 28.9 | 229 |
| Professional services | 245.4 | 4.2 | 46 |
| Other services | 1,185.2 | 18.3 | 97 |
| Communications | 218.2 | 29.0 | 40 |
| Local government | 195.2 | 2.3 | 5 |
| Oil Bases | 0.1 | 0.0 | 0 |
| Oil construction | 25,817.4 | 100.0 | 749 |
| All industries | 32,815.7 | 35.2 | 1,463 |
| All industries (excluding oil construction) | 6,998.2 | 10.4 | 714 |

*Source*: McNicoll and Walker (1978); McNicoll (1980).

exists between output generated by the oil construction sector in other industries and employment in these industries. However the Manpower Services Commission (1981) took the view that his assumptions are unreal as
'one might expect productivity gains to arise through the more intensive use
of labour'. It also believes that the employment multipliers derived by
McNicoll cannot be used to estimate the multiplier impact of the more recent
considerable growth in oil-related employment since his field work was carried out in 1976. If they were deployed now it would exaggerate the employment impact of those developments. On the basis of his calculations he
estimates an employment multiplier for oil construction of 1.85, for oil supply bases of 2.99 and for permanent employment at the Sullom Voe terminal
of over 2. If these were applied to the 1980 oil-related employment figures,
the Manpower Services Commission maintains that over 15,600 jobs in Shetland would either be in oil construction, oil-supply bases or the oil terminal,
or be generated by these industries. Since this total job figure is roughly in

line with total employment in the Shetlands in that year it would imply that all non-oil generated employment had entirely disappeared! Although it may be inappropriate to use these employment multipliers to estimate the absolute levels of future changes in oil-related employment – like the rundown in construction employment – they can still be considered as useful indicators. Thus the actual employment multiplier for supply bases is likely to be higher than the actual multiplier for oil construction whilst the ranking apparent in Table 10.7 of the industries according to the number of jobs created by oil construction may well be of the right order even though absolute numbers may be too high.

To devise more accurate multipliers, the Manpower Services Commission has considered the change in total employment in Shetland (but not including those industries largely unaffected by oil) between 1971 and 1977 alongside the changes in direct oil-related employment. This suggests that an increase in the latter of over 2,500 was accompanied by a rise in total employment of around 3,900 giving a rough multiplier of 1.52. Much of the increase in total employment occurred in the service sector which during this period was in any case rising rapidly elsewhere in the UK. If the rate of increase in this sector had been in line with Scotland as a whole, there would have been around 340 extra jobs, even without oil.

To counter this development there was a fall in employment in the traditional industries (at around 300) which would have a negative multiplier impact on jobs in other industries. This is calculated at 150 jobs, using a multiplier of 1.5.* Taking account of these developments in the service and traditional industries, the 1.52 multiplier might be adjusted downwards to about 1.45. Although fully admitting the crudeness of its approach to deriving employment multipliers, the Manpower Services Commission rightly recognised that it is all that can reasonably be done given the availability of information. Moreover, the methods used are likely to have greater intrinsic merit in an area such as Shetland than elsewhere simply because of the predominant influence that oil has had there on employment trends in recent years.

### 10.6.2 Multiplier estimates – The operational phase

Attention is now given to the important transition from the construction to the operational stage of the most important oil-related aspect of the Shetland economy, the Sullom Voe terminal. Given an overall employment multiplier for oil-related activities of 1.45 and accepting the McNicoll contention that the oil construction multiplier would be lower than that arising from other

---

*Based on employment multipliers for traditional industries quoted in McNicoll, I.H., and Walker, G., 'The Shetland economy 1976/77: structure and performance', *The Shetland Times*, 1978.

*Table 10.8* Possible multiplier effects of construction rundown at Sullom Voe

| Employment multiplier | 1.35 | 1.40 |
|---|---|---|
| Employment with oil industry and Shetland Island Council contractors, 2nd quarter 1980 | 5,669 | 5,669 |
| Estimated multiplier effects | 1,984 | 2,267 |
| of which accounted for by catering industry at Sullom Voe | 1,368 | 1,368 |
| Likely employment loss in other industries | 616 | 899 |

*Source*: Manpower Services Commission for Scotland (1981).

oil-related activities because of the high level of leakage from the local economy, the Manpower Services Commission suggested that the oil construction multiplier might be around 1.35–1.40. If these multipliers are applied to the numbers in oil construction for the second quarter of 1980, then estimates of likely job loss can be derived given the completion of all construction work at Sullom Voe (Table 10.8). If McNicoll's contentions which form the basis of Table 10.7 are accepted then it must be allowed that most of the job losses will occur in the distribution, transport and 'other services' sector, but with an even more significant impact in terms of total numbers remaining employed in the quarrying sector. Apart from the Manpower Services Commission estimates, the Shetland Islands Council's Research and Development Department (1978, 1980) has produced its own figures of the likely multiplier effects of the run down in construction at Sullom Voe. Whilst these suggest a slightly smaller fall than the Manpower Services Commission by 555 jobs, it believes construction work is likely to account for more than 50 per cent of the loss, reflecting the steady increase in non-oil construction employment since the mid 1970s and the production of the McNicoll study. Only transport in the form of road haulage seems likely to experience a substantial fall in employment.

The antithesis of the fall in construction and construction-related employment at Sullom Vow is the creation of permanent jobs in connection with its operation stage. Employment of this kind began in 1978 (at 111) and by the end of 1980 had reached 783. Of the operational interests involved in staffing Sullom Voe the largest employer is BP (47%) followed by the Shetland Islands and Harbours Authority (11%) and Shetland Towage Ltd (10%). Others employed in maintenance, in laboratories, and in catering take up the remaining 32 per cent. Although the total numbers employed in port operations have probably reached their limit, maintenance staff may well rise by about 40 in the next few years. The Shetland Island Council has in addition, estimated a further increase in activities such as diving, plant and vehicle hire, fabrication, office cleaning and banking, thus adding about a further 150 new jobs hopefully to be filled largely by local people who

251

currently occupy just over 40 per cent of jobs. It seems likely that the permanent employment could reach a level of over 900 at Sullom Voe. This in turn will stimulate the employment of local personnel with their expenditure on local goods and services. This spin-off employment resulting from permanent jobs at the terminal should at this stage give rise to a further 200 or more jobs mainly in the service sector.

## 10.7 Regional multiplier estimates

At the broad regional level, no attempts at quantification of the kind used in the Shetland case study are possible, nor could they be applied with accuracy to phases in the development of the oil industry. However, for 1974 an estimate of the numbers employed in oil-related jobs was made by Pounce, Upson and Walker (1976) which gave a total of 26,000 oil-related jobs, excluding those engaged in construction. It was estimated that 50 per cent of these jobs were in new units and 75 per cent in manufacture. Two years later estimates that included construction and offshore installations gave the total oil-related employment at 44,000. If as suggested by the Scottish Economic Bulletin (1977) it was possible to assume a multiplier in the range of 1.2 to 1.4, then total oil-induced employment might be between 55,000 and 64,000. At about this time Mackay and Mackay (1975) were arguing for a general multiplier below 2 while conceding the outside possibility of the oil industry giving rise to a 'super multiplier' capable of initiating substantial expansion. For a region as large and inadequately defined as this must be, notions regarding multipliers must inevitably remain speculative. As Chapman (1976) has argued, they are open to considerable spatial variation. Only a micro study as confined and contained as that of the Shetlands is likely to be close to the truth about the real impact of oil and oil-related industry.

Whatever the total amount of employment directly resulting from oil development, plus those stimulated by it, may ultimately prove to be, its impact on the employment prospects in Scotland as a whole must be regarded as small. The reasons for this may be regarded as two-fold. First is the magnitude of the national problem. This is easily understood when it is realised that according to Mackay and Mackay (1975) between 1966 and 1973 the primary and secondary sectors of Scottish industry were losing 25,000 jobs per year. Moreover in the period 1971–80 in highly populated west-central Scotland up to a quarter of a million job losses were expected. Little wonder they concluded that there was little chance that the benefits from oil would be large enough to produce a significant permanent change in Scotland's economic performance. The second factor is the ultimate local importance of a resource that is easily transportable away to other areas by pipeline. Unlike Odell (1966) who wrongly predicted the transformation of the East Anglian economy as a result of the landing of North Sea gas, Keeble (1976) has been correct in doubting the capacity of energy resources to result in the development of local secondary manufacturing and to have more than minimal impact. Indeed, the product of the oil industry of north-east Scotland mostly passes to Grangemouth into the pre-existing refining and industrial complex. Even here its impact at a national level has been minimal since North Sea oil has largely substituted for imports.

Perhaps it is difficult to disagree with Mackay and Mackay's contention that the prime benefits arising from Scottish oil developments must be seen to be the revenues accruing to the national government rather than direct employment and income creation. These could reach £3.5 billions per year from royalties, petroleum tax and corporation tax by the mid-1980s. Just how much of this might be channelled back into the more general economy of Scotland or that of the north-east region of Scotland remains a matter for political debate and decision.

But the benefits accruing to sub-regions, regions or nations may be but one aspect of change since the introduction of a high-wage industry such as oil with its inevitable influx of specialist workers may disrupt local housing markets, impact on traditional industries by drawing away labour, overload the social services and generally impose strains on established ways of life. In these respects how have those areas of Scotland affected by oil fared?

Where housing is concerned it could be thought that the demands of oil and its related activities might create scarcity in many parts of north and east Scotland, raising prices considerably. In the north the Scottish Information Unit (1975) reckoned that oil needs were responsible for a rise in demand for housing of 10,000 plus units. Of this total the Shetland requirement was probably around 1,200 (Llewellyn-Davies 1975). Assuming an average cost per house of £12,000 (£17,000 in the remoter Shetland area), it was estimated that a total of £120 million needed to be invested to provide all 10,000 at 1974 prices. On the east coast at Cromarty Firth the figure for the period up to 1975 was 4,000 (Hutcheson and Hogg 1974), and for Aberdeen over the years 1972–76, oil and oil-related activities were thought to have created a demand for 16,500 houses (Mackay and Mackay 1975). However, according to Lewis and McNicholl (1978) examining both north and east Scotland, no evidence could be found to show that the house-construction industry could not meet the demands made upon it or that serious shortages were other than of brief duration and fairly isolated. The oil platform construction centre at Nigg was perhaps the most outstanding example of excessive short-term accommodation demand when the rapid build up of the workforce to 1,500 over two years necessitated the importation of two Greek liners to provide living quarters.

As far as the impact of oil and oil-related industries on the traditional forms of employment are concerned, the effects have proved far less dramatic than might have been supposed, according to Lewis and McNicholl (1978). They concluded for rural communities in Scotland's north and east in general that when indigenous basic industries are thriving the loss of la-

253

bour to high-earning oil jobs will only be relatively small. In the case of Shetland in particular they noted that since most of the Islanders seem to enjoy substantial non-monetary advantages from working in these industries, the losses to oil have been particularly modest. Indeed, studies undertaken by the Manpower Services Commission (1981) indicated in each case the much greater significance of external factors to account for movement out of them. Fishing, for example, seemed to have responded only to new patterns of supply and demand as a result of new international 200 mile limits and fishing conservation policies, and any fall in the numbers working in fish processing has resulted from the USA's reduced dependency on imports. Agriculture is essentially of the part-time crofting kind and employment in it remained fairly static over the period of oil-related employment growth. However the knitwear industry (which sustained a considerable element of self-employed home knitters) suffered a 75 per cent drop in its permanent workforce between 1971 and 1977 but this can be almost entirely accounted for by a decline in the fashion market and severe competition from the Far East.

Finally, as far as the impact on the community of oil-based activity is concerned, there is evidence of an influx of oil-related persons straining local social services and relations between these in-comers and the local inhabitants. Nigg, unsurprisingly, proved such an example and in the early days of rig construction, absenteeism, labour disputes and crime were rife with considerable tension between both communities. Of the early years in general Taylor (1974) described a short-term vicious circle of oil development in the remoter areas in which the shortage and expense of available accommodation made it difficult to attract not only the construction workers needed to build new facilities (including housing) but also the professional staff needed to expand the social services as well as health, education and police. However with the perspective of a longer and broader view, Lewis and McNicholl (1978) found it possible to comment favourably on the positive community benefits that have accrued with the development of oil when eventually residentiary workers do move in, bringing with them a requirement for new buildings including schools, hospitals and community centres.

Whatever the impact of oil in north and east Scotland during the exploration, manufacturing, construction, and production phases, what is certain is that at some time – probably early next century – this region will need to come to terms with the depletion of the oil resource. This kind of problem is common to all areas where economic growth has been primarily based on a single mineral resource. As noted in the study cited below by Warren (1973) of the South African Rand area the answer lies in the increasing diversification of the economic base of the region in advance of the exhaustion of its original resource on which its prosperity was founded. Whether this will happen in north-east Scotland, which is in any case locationally highly peripheral to the economic and industrial centre of the EEC, it is far too early to say. However, in the case of the sub-region, Shetland, the passage in 1974 of the Zetland County Act not only gave the local authority powers to acquire development land for the oil industry, but also established a reserve fund from a special tax linked to world oil prices and company profits and levied on all oil entering the island's facilities. Revenue from this should enable funds to be built up which may provide the capital support for the diversification of an already technologically oriented base well before the oil runs out. Taking a wider view, it is possible to argue, along with Lewis and McNicholl (1978) that not only has oil development transformed the economy of the Shetlands and Cromarty, but that there is no reason why the infrastructural investment that has occurred as a result, particularly in transport systems, should not be redeployed in the interests of alternative industrialisation.

## 10.7 Conclusions

What general conclusions may be drawn concerning mining and its capacity not only to stimulate regional development, but to *sustain* its growth in the long term after that original 'mining activity has ceased? Spooner (1981) has identified four key characteristics of resource base development.

1. *The supply of the products of the mining industry within the region must be maintained and the demand for these sustained elsewhere both nationally and internationally in the long run.* The need to remain competitive in cost terms may well be significant internationally though it should be added that uncompetitive regional production might be maintained via the use of tariff barriers to keep out international imports. Mining areas which must operate in a competitive market but cannot of themselves remain cost-effective or which rapidly exhaust their richer deposits are not likely to be of value in the promotion of regional development.
2. *The large scale of the operation to exploit the resource in question is a key factor and is frequently allied to a large labour force.* Any analysis of coal field developments from the eighteenth and nineteenth centuries in Britain indicated these as fundamental to their on-going importance. Warren's work (1973) on the Rand which traces its development from a goldfield to the largest industrial agglomeration in Africa also underpins such a concept. From the discovery of gold there in 1886 and the establishment of a mining camp, it had grown in twenty years to a city with a population of 150,000. Energy needs were met locally when coal was found close by at Witbank and local agriculture was easily able to support the requirements of the growing community. Crucial to the ultimate success of the region as one of sustained growth, and to the region's major role in the economic development of South Africa as a whole, has undoubtedly been the scale of the mining industry in the Rand.

3. *The development by the mining industry of both forward and backward linkages may also be significant aspects of its capacity to create and sustain regional growth.* Richardson (1976) recognised the need for its complementarity with other activities. Spooner (1981) cited the strength of the coal industry in Britain in the nineteenth century and its capacity to draw user industries to the coalfields. Earlier in this book the role of beneficiation, smelting and basic fabrication was noted particularly where the ferrous and non-ferrous metals are concerned, as well as the increasing importance these may have in terms of economic growth in the less developed countries.

4. *If it is to play a key role in regional development, a mining industry must not be characterised by a high level of 'leakage' in terms of its own use of goods and services.* Spooner's (1981) fourth point here is perhaps a key one given the emphasis placed in this chapter on multipliers and their impact. In both the case studies of tungsten and oil, each widely separated in so far as its propensity to create development locally is concerned, the importance of local operational expenditure has been recognised not only in terms of the mining operation's own material or non-labour needs and its direct employment of labour, but more pertinently in its capacity to create additional jobs in the residentiary sector of the economy.

In concluding, Spooner (1981) added a footnote in which he stated that the region must become attractive to migrants. This point is worthy of development into a fifth key characteristic of particular concern to both the public at large and industrialists in the late twentieth century. Even if it can be accepted that, say in the case of coalfields, the ubiquity of the power they now provide means they are less of a regional draw than in the nineteenth century, their infrastructural advantages ought to allow them to remain attractive. The fact that they are not of sufficient significance to footloose entrepreneurs who now make up a large sector of the industrial base of many of the developed countries of the world, when weighed against the environmental disadvantages of these 'old' industrial regions, is borne out in a number of studies. As early as 1968 a survey of management in the Northern Planning Region of Britain (House *et al.* 1968) recognised that the 'poor' or negative image of many of the northern coalfield areas was a major factor in pushing new industrial development to the environmentally more favoured south. A year later the Hunt Committee Report in the UK also recognised the lack of attractiveness of the environment of old mining areas, advocating government grants for the reclamation of derelict land so as to enhance their appeal to potential industrialists.

The attraction of new industries, the products of which incur only low unit transport costs, to more environmentally attractive locations was recognised in a number of studies in the early 1970s (Townroe 1971; Blunden 1972; Green 1974). Keeble (1976) wrote that 'the image of particular localities as attractive residential environments is now in the 1970s a fact of major importance in manufacturing growth strongly influencing decisions of both workers and industrialists'. As for the future possibility of mineral extraction as a basis for regional growth, it is evident that within the developed countries, particularly Western Europe and North America, adherence to high environmental standards will be a legal pre-requisite of any attempt to establish such enterprise. In the latter area the importance of the Environmental Impact Statement as a means of ensuring minimal damage to the landscape and to ecosystems at the development and production stages of mining, as well as after the deposit has been exhausted, cannot be underestimated.

In the UK context the research group set up by the government to investigate the likely impact of mining on the environment has provided useful guidance to the minerals industry and to the local authorities responsible for the evaluation of development proposals which, if adhered to, should mitigate problems in the future (Blunden *et al.* 1973). In the longer term the replacement of medium and small size hard rock quarries by regionally or nationally based operations of a very large size (over 10 million tonnes of valuable material per year) and remotely located was postulated by the group, along with the underground working of some such materials (particularly the higher grade limestones). But as a more immediate and pragmatic response, basic and inexpensive practices regarding screening and landscaping to ameliorate the impact of open-pit operations were put forward together with the idea that regionally based advisory teams should be created and staffed by minerals planners, mining engineers, landscape architects and so on, to assist local planning authorities in their decision-making regarding minerals development. The notion of regional advisory teams was also suggested in evidence given by the research group to the Stevens Committee (1976) on the reform of minerals planning law and it was indeed recommended in the final report of the latter. If and when these suggestions are acted upon in the UK context, it should help to temper the reaction to potential mining operations by vociferous environmental pressure groups at numerous public inquiries. Since the early 1960s, such groups have maintained that the landscape may be irreparably changed in the interests of a transitory activity whose value to the economy of the area may be dubious.

In the meantime attempts to do more than merely address levels of direct employment as a means of stressing the positive impact on an area of mining activity through the quantitative analysis of multiplier effects may be helpful in the justification of such projects.

# CHAPTER 11

# The long-term availability of minerals

## 11.1 Alternative approaches

When a mineral deposit is located at a specific site, its working potential can be estimated by a detailed programme of exploratory drilling. However, when it comes to forecasting the likely capacity of the earth to meet future mineral needs, serious difficulties arise. Apart from the problems regarding demand already identified, estimates of supply are hampered by only partial knowledge of the possible locations of mineral deposits. When these are known, the doubtful accuracy of averages worked out without the aid of detailed exploratory investigations for highly variable concentrations of minerals, especially metallic ores, make for added uncertainty. The profitability of working deposits of a given size and grade at a particular location together with their practicability in terms of the technology available are further variables, the significance of which may change over time. In spite of all these difficulties, the question of the likely future long-term availability of resources, particularly of the more valuable kind, has received increasing attention especially in the last few years in which resource conservation in the widest sense has been a matter of popular debate. Two strongly opposed schools of thought appear to have emerged. First, there are those who take a pessimistic view of the earth's capacity to meet the demands made upon it. They see a contradiction between present behaviour and the future expectations of society. Their argument is that, given the likely increase in population and greater demand from the less developed as well as industrialised countries, continued resource exploitation will lead to untenable levels of pollution. In the long run there will also not be sufficient mineral resources to sustain anything which resembles the way of life extant in the developed world. They predict the collapse of society, a prediction based on quantitative forecasting using past and current data. A contrary and more optimistic view is taken by their opponents who note a wide-ranging number

258

of precedents as evidence for the likelihood of new technological remedies being made available to avert such disasters.

Analysing the views of the pessimists in greater detail it is apparent that their predictions rest largely on an application of systems theory (an extension of the laws of thermodynamics) and on the phenomenon of exponential rates of growth.

The first law of thermodynamics, or the law of conservation of energy, states that energy – the capacity to do work – cannot be destroyed. It implies that energy can only be redistributed or, to use systems theory language, transduced. To take a simplified example: the energy in food is a capacity to do work which is realised when a man eats it. His ability to burn or metabolise that food provides him with such things as mechanical energy and the ability to move. This in itself increases respiration and the release of thermal energy to the atmosphere. Any energy redistribution requires the three components illustrated in this example: first, a pool of energy resources (in this case, food) which can be termed inputs to the system; second, a transducing agent (in this case, man) capable of realising this energy by converting it into work; and third, an environmental sink (in this case, the atmosphere) capable of assimilating the redistributed energy, which may be termed the outputs to the system and many of which come under the blanket term 'pollution'. Together these three components make up a system which will continue to function as long as the components are in equilibrium. Disequilibrium caused by continuous stress at any point on the system, will lead to its eventual collapse.

Figure 11.1 represents the application of systems theory to the world economy, with fuels, food and raw material as the inputs, human population as the transducer and the environment as the output sinks. Overloading or stress can originate only from the transducer. His intervention is required to realise both the potential of the energy resources – by converting them into work – and that of the sinks as receptors of outputs.

Those adopting the pessimistic approach maintain that an overloading of the world economy system has been taking place since the 1960s due to the increasing demand for inputs. This demand is itself a symptom of the rising material expectations and rapid growth of population and, it is maintained, can only emulate these by growing at an exponential rate. It represents one aspect of the problem as they see it, namely resource depletion. The other arises as a direct consequence, for the overloading of demand for inputs means an increased amount of energy redistribution and a corresponding overloading of the sinks with outputs or pollution.

The second pillar of the pessimist argument is exponential growth, really a restatement of the nineteenth century Malthusian theory of the geometrical progression of population. It is best explained by reference to doubling times. The subject under study usually progresses through several doubling times without appearing to either reach a significantly greater size or in-

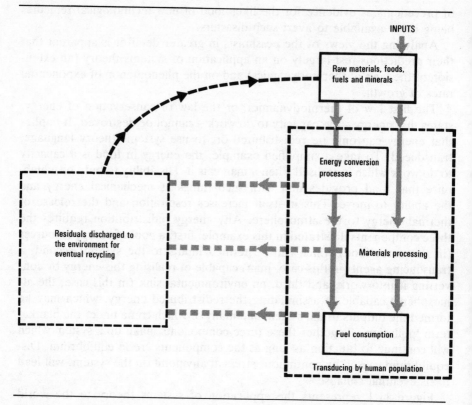

INPUTS

Raw materials, foods, fuels and minerals

Energy conversion processes

Residuals discharged to the environment for eventual recycling

Materials processing

Fuel consumption

Transducing by human population

*Fig. 11.1* A systems theory approach to the world economy. *Source*: Blunden (1977)

creasing its rate of growth very much. Beyond a certain threshold, however, it suddenly surges to reach overwhelming size within a very few doubling times, perhaps one or two only. A simple example illustrates this: imagine a pond with a waterlily growing on it which is doubling in size every day. If it is known that the waterlily will completely cover the pond in thirty days then it demands very little reflection to forecast that it will not cover half of it until the twenty-ninth (Brown 1978).

Exponential growth also applies to the dynamics of human population increase. Projections of population size as far back as 8000 BC can be computed from such contemporary census figures as survive and from the deductions of archaeologists about the size of settlements. These indicate that in the 9,650 years from 8000 BC to AD 1650 the population doubled between six and seven times, from five million to 500 million, an average of once every 1,500 years. Any estimate of the rate at which doubling time was reduced would be hypothetical and largely irrelevant insofar as it must have taken far longer for the population to double from 250 million to 500 million than the subsequent two centuries it took to reach 1,000 million by 1850. Population size in 1650, therefore, represents the threshold in its exponential growth: it took another 280 years only for it to quadruple to 2,000 million in 1930 – the second, greater doubling from 1,000 to 2,000 million occurring after 1850.

At present rates of growth the world's population, having reached 4,000 millions in the mid-1970s, a reduction of thirty-five years over the previous doubling time, will double again to reach 8,000 million in the subsequent thirty-five years. Steps to halt world population growth are held by the pessimists to be even more urgent in view of the time lapse which would occur before they took effect due to the world's predominantly young age structure.

Thus population growth would be one major cause of the increasing demand for inputs, principally raw materials, among which are the metallic and non-metallic minerals. Indeed Dunham (1978) has spoken of the way in which, over the past 120 years, 'the curves of mineral consumption shadow the population curve' (Fig. 11.2). But another significant demand factor must be the switch in the less developed economies to a capital-intensive, consumption-oriented economy. However it may be expressed, it is the aim of the overwhelming majority of people to improve their standard of living as measured by the consumption of inputs. A country's gross national product is necessarily a measurement of input consumption as well as output and it is the policy of governments to achieve rates of economic growth well above the rate of population increase to ensure continuing improvements in living standards. Most EEC countries, for example, achieved economic growth rates of over 7 per cent per annum in the years between the mid-1950s and 1973, while population increase averaged 0.8 per cent. Also, up to 1973, world consumption had been growing at around 4 per cent a year at a compound rate, which meant it was doubling every fifteen years. In other words it was growing at least twice as fast as population, a trend which, by shortly into the next century, might have led to a population that could well be approaching twice its present size, demanding four times the levels of inputs extant in 1973.

Though it is now clear that the world recession of the late 1970s and early 80s will leave demand short of such predictions, it has not ceased to validate (as they see it) certain of the gloomy predictions made by the pessimists concerning the depletion of non-renewable inputs such as fossil fuels and minerals. They contend that this period of recession is only a temporary downturn (albeit triggered off by the oil crises) in an indentifiable economy cycle (Forrester 1976) from which the industrial nations will recover by the late 1980s.

Citing as an example the estimates of world oil reserves and their forecast rates of consumption (as in Chapter 3), the pessimists would maintain that oil supplies will be exhausted by the last quarter of the twenty-first century. More interesting, from the point of view of emphasising the nature of ex-

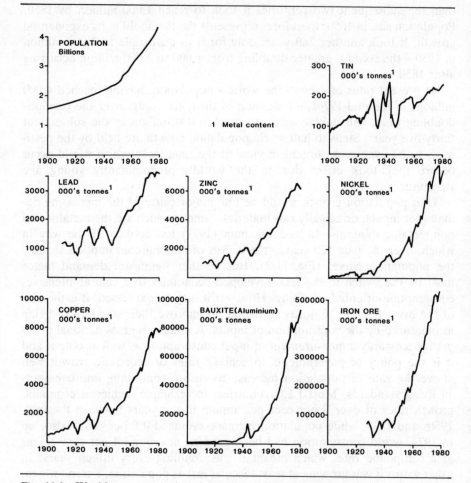

*Fig. 11.2* World consumption of certain key minerals compared with population growth from *c.* 1900. *Source*: United Nations

ponential growth of demand, only 12 per cent of all the world's oil was used by the end of 1975. Similar predictions forecasting the exhaustion of reserves of the principal metalliferous ores, with the exception of chromium and iron, within fifty years were also made prior to the oil crises of the 1970s by a principal exponent of the pessimistic view, Thomas Lovering (1969). Others in the same camp as Lovering, including Meadows *et al.* (1972), have taken only a slightly less gloomy view (Table 11.1).

Whatever credence is given to their views, at the rates of consumption prevailing at the time of the Meadows *et al.* predictions, it would have taken only another thirty years to consume an amount of metal approximating to the total used in all previous time. Pessimists contend that such economic growth rates and the policies and philosophies that support them are both

*Table 11.1*  Forecasts of mineral exhaustion according to Meadows *et al*.

| Mineral | Number of years supply under existing rate of use (1970) | Number of years supply assuming continued growth in rate of use | Assuming reserves are 5 times those now known | US per cent of total world consumption |
|---|---|---|---|---|
| Gold | 11 | 9 | 29 | 26 |
| Mercury | 13 | 13 | 41 | 24 |
| Silver | 16 | 13 | 42 | 24 |
| Tin | 17 | 15 | 61 | 22 |
| Zinc | 23 | 18 | 50 | 26 |
| Copper | 36 | 21 | 48 | 33 |
| Lead | 26 | 21 | 64 | 25 |
| Tungsten | 40 | 28 | 72 | 22 |
| Aluminium | 100 | 31 | 55 | 42 |
| Molybdenum | 79 | 34 | 65 | 40 |
| Manganese | 97 | 46 | 94 | 14 |
| Platinum | 130 | 47 | 85 | 31 |
| Nickel | 150 | 53 | 96 | 38 |
| Cobalt | 110 | 60 | 148 | 32 |
| Chromium | 420 | 85 | 154 | 19 |
| Iron | 240 | 93 | 173 | 28 |

*Source*: Meadows *et al*. (1972).

unsustainable and indefensible in the face of exponential resource depletion. They advocate a completely new set of definitions by which to judge living standards, based on qualitative considerations in which other non-material criteria would be paramount beyond a certain level of wealth. This level would be fixed at the conversion point from an open-flow economic system to a closed, stock one. This is an idea which Boulding (1971), an economist, referred to when he contrasted the cowboy's mentality with his assumption of limitless frontiers to the spaceman's recognition of the strictly circum-scribed limits imposed by the size and life support systems of his spacecraft. In a closed, stock economy, equilibrium would be imposed, the supply of inputs being limited to the technological and environmental sink capacity to recycle outputs. Pessimists believe that a step towards implementing such a programme would be the imposition of a raw material tax inversely pro-portional to the availability of the raw material and the demand for it, and an amortisation tax tied to the estimated life of a product.

This drive towards resource depletion is compounded, the pessimists maintain, by the economists' traditional assumption of the environmental sinks as free areas for the dumping of outputs (Hardin 1978). In this situation problems of allocation arise which cannot be solved by the price mechanism since the sinks become overloaded. Consumer preference, directed by price, cannot determine the optimum use of the environment as a receiver of wastes, either because society is unwilling to curb pollution – seeing it as the cost of material well-being – or lacks the knowledge to attribute it to particular decisions.

On the other hand, optimists (Beckerman, Maddox, Zuckerman, *et al.* 1972), point out that while systems theory is indisputable, being based on unequivocal physical laws which have been proven time and time again, its blanket application to the complicated sphere of demand and supply interactions is tendentious. Whilst accepting the contention that continued population growth at the present rate is a serious matter which should engage the urgent attention of humanity, they maintain that predictions, based on the phenomena of exponential growth insofar as the consumption of minerals is concerned, take no account of how society reacts to actual or expected increases. The optimists' argument hinges on the viability of basic economic philosophy. They say the changing patterns of demand and supply as determined by availability expressed by price will allocate resources in a situation of resource depletion and pollution. Thus, whereas their opponents counsel conservation of dwindling resources, optimists reply that real or expected shortages lead to price increases. These in turn lead to pressure for technological innovation which may take one of three forms.

First, such innovation may lead to new methods of finding, processing and/or making use of minerals previously not considered part of the resource base let alone considered as part of the reserves. For this reason alone it is erroneous to extrapolate into the future as Meadows *et al* have on the basis of USBM data as if resources and reserves were immutably fixed. A good example of the way in which the resource and then the reserve stock of minerals may be added to by technological advance is that of ocean mining. Although the existence of nodules on the seabed has been known for a hundred years (Enzer 1980), the mining of these did not enter into the considerations of Meadows *et al*. But here is an example of the economic mechanisms already cited producing technological development to the point at which the relevant hardware for harvesting these nodules has become available. For operations to commence, expected market prices must improve substantially on the recessionary levels of the early 1980s and/or harvesting costs will have to fall dramatically. Even in the shorter term relatively modest levels of nodule harvesting will impact on world supplies and prices of manganese, cobalt and nickel but to a much lesser extent, copper. This is because of the percentage content of the individual minerals in the nodules in relation to overall world consumption (Table 11.2). Nevertheless in the long term their identification as resources and translation to the status of reserves represents an important facet of the optimists line of argument.

The second technical innovation concerns the greater 'spreading' of the mineral reserves available. Such spreading may occur in the manufacturing process or it may make itself felt through the second course open to technology – the recycling or recovery of wastes.

Third, technology may come up with substitutes. Of these last two possi-

*Table 11.2*  Impact of different levels of seabed mining in 1990 and 2000 under low and high rates of mineral demand

|  | *Nickel* | *Copper* | *Cobalt* | *Manganese* |
|---|---|---|---|---|
| | (Thousand tonnes) | | | |
| Projected world consumption | | | | |
| 1990 (low) | 1,054 | 13,084 | 43.4 | NA |
| (high) | 1,380 | 14,478 | 48.0 | |
| 2000 (low) | 1,314 | 16,839 | 57.5 | 19,400 |
| (high) | 2,357 | 21,330 | 77.5 | |
| Production from seabed | | | | |
| at 3 million tonnes | 35 | 30 | 5 | 630 |
| at 10 million tonnes | 116.7 | 100 | 16.7 | 2,100 |
| at 15 million tonnes | 175 | 150 | 25 | 3,150 |
| | (Per cent) | | | |
| Seabed share of market under low (high) demand | | | | |
| *1990* | | | | |
| at 3 million tonnes | 3.3 (2.5) | <1 | 11.5 (10.4) | N/A |
| at 10 million tonnes | 11.1 (8.5) | <1 | 38.5 (34.8) | N/A |
| at 15 million tonnes | 16.6 (12.7) | 1.1 (1.0) | 57.6 (52.1) | N/A |
| *2000* | | | | |
| at 3 million tonnes | 2.7 (1.5) | <1 | 8.7 (6.5) | 3.2 |
| at 10 million tonnes | 8.9 (5.0) | <1 | 29.0 (21.5) | 10.8 |
| at 15 million tonnes | 13.3 (7.4) | <1 | 43.5 (32.3) | 16.2 |

*Source*: Saunders (1980).

bilities something more now needs to be said because although both are practised and were commented upon in Chapters 5, 6 and 7, ultimately both can enjoy a much greater potential.

As far as recycling is concerned, Table 11.3 summarises the performance of one nation, the USA, in terms of its reuse of metals. Whilst it is perhaps representative of other developed countries in the late 1970s – though there are some differences – it is by no means indicative of what could be achieved at some future date to raise the current levels of mine-produced metals recycled (25–30% according to Walker 1979) and this in spite of the relatively high percentage of copper and crude steel currently being so treated. Apart from active measures to encourage recycling, further technological developments could be of considerable significance. Steel-making furnaces are known to give off oxide effluents from which zinc and lead might be recovered. According to the United States Bureau of Mines, 100,000 tonnes of zinc and 10,000 of lead could be retrieved annually from furnace dusts once an appropriate methodology emerges from its ongoing research programme (Henstock 1975). However, one major stumbling block in recycling is the metallurgical complexity of many pieces of redundant household and business equipment. Within these, copper and high silicon steel components

*Table 11.3*  Recycling metals (USA)

| Metal | Approximate annual recovery from scrap* | Remarks |
|---|---|---|
| Iron | 71 to 80 million tonnes† | Includes both new (about 60%) and old scrap. In the iron cycle from mine to product to recovery, the loss of iron is 16 to 36%. About half the feed for steel furnaces is scrap. |
| Ferro-alloy | | Most special purpose steel scrap containing a high percentage of chromium, cobalt, molybenum, nickel, tungsten, and vanadium (high temperature alloys, stainless steel, high-strength and high-speed steel) is re-melted without separation of the alloy metal. Tungsten is recovered from carbide and metal sludges and recast into carbides. Manganese is recovered by chemical and metallurgical industries from spent $MnO_2$ oxidising solution. Over 11,000 tonnes per year of nickel are recovered from non-ferrous scrap. Recovery of molybdenum from non-ferrous scrap is small. Over 12.0 million tonnes per year of manganese rich open-hearth slags are produced in the US and some 25% of them are re-cycled; these slags contain 5 to 9% manganese and are potential sources. Some slags from re-melting of steel contain up to 12% chromium but these are not currently used as sources. |
| Copper | Over 1 million tonnes | Secondary copper production from old and new scrap ranges from 914,000 to 1,016,000 tonnes per year, about half of which is old scrap. Old scrap reserve is estimated at 35.6 million tonnes in cartridge cases, pipe, wire, auto radiators, bearings, valves, screening, lithographers' plates. 20% of consumption is from recycled copper. |
| Lead | Over 0.5 million tonnes | Estimated reserve is 4.1 million tonnes of lead in batteries, cable coverings, railway car bearings, pipe, sheet lead, type metal. Over 44% of consumption is from recycled lead. |

| Metal | Approximate annual recovery from scrap* | Remarks |
|-------|----------------------------------------|---------|
| Zinc | Over 0.25 to under 0.5 million tonnes | Zinc recovered from zinc, copper, aluminium, and magnesium-based alloys (about 14% of zinc pool is recovered representing 7% of consumption). |
| Tin | 20,300 to 25,400 tonnes | Tin recovered from tin plate and tin-based alloys, 20%; tin recovered from copper and lead based alloys, 80%, aking 21% of consumption. |
| Aluminium | Over 0.3 million tonnes | Because aluminium is a comparatively new metal the old scrap pool is small, but it is growing rapidly. (About half the available pool is reclaimed and represents about 10% of consumption.) |
| Precious metals | gold = c.30 tonnes silver = 850 tonnes | Precious metals including platinum are recovered from jewelry, watch cases, optical frames, photo labs, chemical plants. Because of the high value, recovery is high. |
| Mercury | 365 tonnes | Recovery is high. Nearly all mercury in mercury cells, boilers, instruments, and electrical apparatus is recovered when items are scrapped. Other sources are dental amalgams, battery scrap, oxide and acetate sludges. |

* Includes old and new scrap. New scrap or 'home' scrap is produced in the metallurgical and manufacturing process: old scrap is 'in use'.
† Old and new scrap consumed rather than recovered.

*Source*: Adapted and updated from Flawn (1966).

(materials which have considerable scrap value) are frequently in close association in components such as motor armatures, field cores, transformers and so on. Here again the United States Bureau of Mines is working on methods of disassociating the two based on preferential melting in molten salt baths (Henstock 1975), a technique which also may assist in the separation of the constituent metallics of ferro-alloys (Table 11.3).

As well as future technological solutions, governments may play a part in encouraging recycling and thus the conservation of minerals. In this respect the tipping at sanitary land-fill sites of wastes containing metallic materials ultimately may be forbidden whilst the mandatory sorting of waste materials at the household or factory end of the disposal chain could assist in the reduction of recycling costs. However, Butlin (1977) has made a number of suggestions to lower the 'take' of new minerals and to encourage recycling including the levying of a tax on the use of 'virgin' minerals for certain manufacturing processes and the inclusion in the selling price of a consumer durable, such as a car or refrigerator, a sum which is refundable when the product is no longer usable and is returned to a centralised collecting centre for recycling. Conn (1977) has suggested subsidies for retailers of beverages who wish to convert to the handling of their product through reusable containers, an approach which also has the added advantage of considerable savings in energy as a Canadian study (Solid Waste Task Force 1974) has already demonstrated. Indeed, recycling in general may be regarded as involving a far lower level of energy consumption than that involved in the winning, processing and smelting of newly won metallic ores. For instance, it takes only 5 per cent as much energy to melt down old aluminium as it does to produce virgin metal from ore, and remelt furnaces are comparatively inexpensive to construct.

Recycling will always involve some loss of material and Solow (1974) has likened the process to that of the multiplier in macroeconomic theory. In examining the immediate potential for copper recycling in the USA, Banks (1977) suggested that an additional 10 per cent of total consumption could be obtained in this way. This was in line with the calculations of Fischman and Landsberg (1972) who in evidence to the US Commission on Population Growth and the American Future were at that time postulating that a further 25 per cent of the aluminium requirement in the USA could be obtained by recycling, along with an additional 18 per cent for zinc, 3–4 per cent for lead, and less than 2 per cent for iron.

A good deal was said about the substitution of materials in Chapters 5, 6 and 7 on individual minerals. Goeller and Weinberg (1976) argued at a somewhat theoretical level that most minerals can be substituted in various ways, and Rajaraman (1976) offered a list of substitutes for tin and lead. Problems arise over the fact that the substitute for lead is zinc and the two are so frequently found together that some mineral analysts treat them together (Dunham 1974). Substitution is consequently constrained by the risk that both would be scarce at the same time. Roberts (1973) has suggested that this consequence could also arise from a higher degree of interdependence between copper, lead, zinc and tin. However, plastic materials made from tough polymers are now increasingly appearing as substitutes for metallic materials. Whilst it is difficult to quantify the exact impact of such a substitution there is evidence from the automotive industry that more stringent safety standards and the desire to produce lighter vehicles to improve fuel consumption, led in the period between 1975 and 1980 to the replacement of 454 kg of metals by 204–17 kg of plastics. The possibilities here are by no means exhausted.

All these considerations of recycling and substitution (as well as those of the discovery of further deposits of particular resources and their possible translation to the status of a reserve) are within the realms of real world experience and practice. It is in this respect that the optimists would claim superiority over their opponents whose approach, in their view, is largely theoretical. But in order to crystallise the claims and the attitudes on both sides of the argument regarding the long-term availability of minerals, we can do no better than consider the views of the optimists and the pessimists in relation to two case studies.

## 11.2  The supply of oil: a case study

Examining the prediction of exhaustion of oil in the twenty-first century, optimists attack not the assumptions about levels of future demand so much as the failure to consider the likely reactions by the oil industry and governments. It is pointed out that the doubling of oil consumption every decade up to 1973 and the need to diversify supplies thereafter improved exploratory techniques which have forced an upward assessment of total resources. In 1942 world resources were estimated at $600 \times 10^9$ barrels, but by 1978 estimates ranged as high as $4,500 \times 10^9$ as was noted in Fig. 3.7, Chapter 3. This confirms the influence of extended exploration into the polar regions, offshore, and to greater depths into rock. Such new discoveries are, of course, the easiest options open to technology in its solution to the demand problem.

Optimists also predict that technology will react to the knowledge that oil is a finite resource and to rising prices, as conventional oil resources are depleted, by the development of substitutes. So the interplay of shortage and price will be seen in the sequential or simultaneous development of new sources of oil and substitutes for oil.

The speed at which these developments take place will be partly a political and not a purely economic decision. Nations will not feel secure if they depend for a major source of energy on imports from countries with political ideologies opposed to their own. The force of this argument was brought home to a number of countries whose attitudes to the Arab/Israeli conflict of the 1970s had to be changed to ensure oil supplies were continued, even at a reduced rate. Moreover, one country with similar political ideologies to another may now be no longer in a position to ship oil as part of a mutual trading pattern because domestic demands must take precedence. For example, in the crisis following the same Arab/Israeli war, Western European countries did not receive oil from the USA, something which had occurred in the 1956 Arab/Israeli conflict.

Among the new sources of oil that will be turned to are so-called synthetic oils contained in oil shales, tar sands and coals. The technology to extract oil from these materials already exists but their uncompetitive price as a refined product acted as a brake upon their exploitation. However, in the case

269

of the Alberta tar sand deposits, by early 1979 extraction prices were not a great deal more than those of North Sea, as Chapter 7 made clear, and even closer to coinciding with the increasing costs involved in discovering and developing new fields in the United States, with the exception of Alaska. It is also possible that some more efficient method of exploiting conventional oil fields may be discovered. At the moment up to 50 per cent of the oil present in a field remains in the rock. The development of mining technology to recover this residue cannot be discounted. Indeed, research carried out by Moses (1983) suggests that by using microbial technologies oil well yields could be trebled. Such biotechnology would be used to produce mobilising chemicals designed to flush out the oil more effectively, to act as an agent for the direct mobilisation of residual oils, or as a means of isolating oil reservoirs to prevent loss by seepage. Potential supplies of oil computed from known deposits of oil shales, tar sands and coals are enormous, and from the point of view of the industrialised nations favourably distributed. However, it is just possible that even before they are developed to any significant degree, substitutes might take the place of oil.

Oil in 1975 occupied the place that coal did in 1875 as the foundation of industrial production. Appliances and processes are designed to run on oil – a limiting factor in the short run but one which can be nullified more easily and quickly than is usually realised. Although oil is the principal energy source of modern industrial society, the greatest demand for it comes from commerce, industry and public transport systems, precisely those sectors over which cost-conscious companies and governments have the greatest control – rather than from individuals for private use. As the cost of conventional or synthetic oil rises, the responsibility and pressure on governments and industry to substitute other and in some cases new energy sources for it will also increase as will their capacity to use it more efficiently.

Although the world's economies have grown by 20 per cent in real terms, since 1973 energy consumption has only risen by under 4 per cent, and the amount of oil burned has actually declined. Of the substitutes for oil, Table 11.4 indicates the likely development potential of each, leaving aside the possibilities for another dwindling resource, coal, which must be looked upon as a fuel which will effect an energy bridge between the current reliance on oil and the energy sources of the future. Table 11.4 shows that the almost certain outcome must be higher levels of research into nuclear energy sources leading to a choice of the most cost-and-production-efficient reactor and a resolution of the radio-active waste problem, leaving oil to make its contribution only in the transport sphere and petro-chemical industry. Mass transport systems utilising other energy sources are already well-developed and should the private car remain socially viable beyond the end of the century an alternative fuel would presumably be found for it. The demand for oil from the petro-chemical industry is relatively insignificant beside these other uses. It has been estimated, for instance, that to switch the material for all car bodies from steel to plastic would – at current pro-

*Table 11.4* The magnitude of new energy sources

| | World annual capacity in $10^3$ megawatts | | | | |
| | Solar | Hydro-electric | Tidal | Geothermal | Nuclear |
| --- | --- | --- | --- | --- | --- |
| Current (c.1980) | 0.1 | 152 | 0.1 | 1.12 | 617,855 |
| Probable maximum | Small-scale special purpose | 2,860 | 64 | 60 | Large but highly dependent on technological developments |

*Source*: Blunden (1977).

duction rates – increase the demand for oil by less than 1 per cent (Blunden 1977).

It is then perfectly possible to argue from the current situation that by the end of the century society will be suffering from a severe oil shortage, or that the commodity will be in the same situation as coal up until the autumn of 1973 with more than adequate reserves for which there is a falling demand.

## 11.3 The supply of metals: a case study

Under the stimulus of demand, exploratory techniques have increased the supply of metallic minerals and made the mining of lower grades of ore economically viable as the richer deposits are worked out. Optimists see no reason why this state of affairs should not continue. They therefore reject the notion of Meadows *et al.* (1972) that in making forecasts it is possible to consider the rising demand for minerals within the context of proven reserves as if they represented something that is fixed and immutable.

Pessimists hold that the situation is not as clear-cut, being complicated in detail by geological factors and in general by a misinterpretation of Lasky's work (1950, 1955). Lasky expounded the AG (arithmetic/geometric) ratio to determine the percentage of rock to ore grade with reference to copper in the United States. Averaging the characteristics of the major eight copper mines in the country he forecast a geometric increase in tonnage of mineralised rock as the grade of content of copper decreased arithmetically. In other words, the leaner the copper ore the greater the amount of ore. As Lasky stressed, this is not equivalent to saying that the leaner the copper ore the greater the amount of ore and, therefore, metal available. Lasky's average copper mine covered a range of grade of only 1.5 per cent and exhibited an 18 per cent increase in tonnage of mineralised rock for each decrease in grade of 0.1 per cent. Nevertheless, he calculated that the vast bulk of copper would be won from the ore above 0.5 per cent, a third of it from 175 million tonnes with a grade of 0.5–0.9 per cent, and over a half from the even smaller volume of rock containing more than 0.9 per cent copper. Pessimists argue that those who confidently predict the future availability of metals, basing their predictions on the AG ratio, have misread Lasky, failing both to note that he was dealing with one particular metalliferous ore within a very limited range of grade and failing to draw the full implications of the ratio for the size of operations.

Several metals do have an average abundance which is not a great deal less than the present cut-off grade of ore, below which it is unprofitable to mine. Iron ore usually contains 20–30 per cent iron, while the average abundance of iron in many rocks is around 5–8 per cent. However, as the Russians have already profitably exploited blacksands with 4 per cent iron content it is foreseeable that iron could be mined at its average abundance in the future. Bauxite, cobalt, nickel and vanadium also exhibit relatively small ranges of grade between present cut-offs and average abundance, while in many igneous rocks the copper content is several hundred parts per million – only one order of magnitude below the present cut-off of around 0.4 per cent.

Ore and average abundance grades of some of the other metals are, however, vastly dissimilar. Metals such as mercury and tungsten are concentrated by geological factors, such as volcanic activity, chemical leaching along fissures and so on, into rich deposits sharply segregated from the surrounding rocks. Mercury, for example, comes from a few major deposits with a grade of 0.2–0.5 per cent, whereas the average abundance of mercury is 0.00004 per cent – four orders of magnitude below the present cut-off.

Such large ranges between present ore grade and average abundance raise two questions. First, presuming the technology were available to utilise rock at these very low grades, how could it be profitably worked? Lovering (1969) has used United States Bureau of Mines statistics covering five decades up to the 1960s to demonstrate that copper mining costs in that country are no longer falling, but have levelled out. This evidence was used by Lovering to counter the arguments of Barnett and Morse (1963) that technology would stay ahead of demand and be reflected in falling costs.

Copper is one case where the present ore grades and average abundance are not widely different. If technological progress cannot absorb the increased costs of working the leaner copper ores presently entertained, how can it be expected to overcome the financial obstacles to working at the average abundance level? This question becomes even more pertinent when mercury and other metals with an average abundance hundreds or thousands of times leaner than the present ores are considered.

Scale of working is a dominant consideration from a technological cost viewpoint as well as that of the environment. To take again the case of copper: mines such as Bougainville in the Solomon Islands and Bingham, Col-

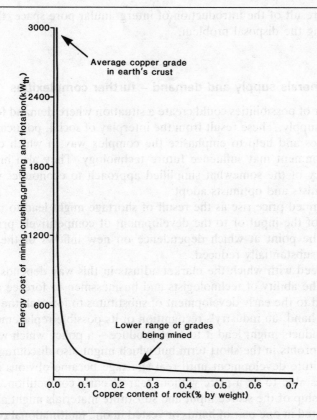

*Fig. 11.3* Energy requirements for mining, crushing and flotation of different grades of copper sulphide bearing rock. *Source*: Singer (1979)

orado, are enormous undertakings but only one-tenth the size of operations which would be required to yield comparative tonnages of metal from rock an order of magnitude leaner. Energy is another equally dominant consideration. As Fig. 11.3 illustrates, huge increases in the expenditure of power are required in the recovery of copper in moving away from currently mined grades to approach those of the average copper grade in the earth's crust. As to whether such an input is possible, much will depend on the future availability of cheap and plentiful sources of power perhaps generated from nuclear sources. But for the gigantic consumption of both land and energy one would have to turn to the working of mercury or similar such metals at or near their level of average abundance. Here the implications are really quite unthinkable with, as a corollary of the AG ratio, a geometric increase in the tonnage of waste rock to be disposed of for every arithmetic decrease in grade. It also has to be remembered that for every tonne of waste rock removed from the ground, its volume will increase by between 20 and 40 per

273

cent as a result of the introduction of intergranular pore space, thus further aggravating the disposal problem.

## 11.4 Minerals supply and demand – further complexities

A number of possibilities could create a situation where demand for minerals outstrips supply. These result from the interplay of social, political and economic forces and help to emphasise the complex way in which concern for the environment may influence future technology. They also highlight the inadequacy of the somewhat simplified approach to economics which both the pessimists and optimists adopt.

A sustained price rise as the result of shortage might lead to recycling or recovery of the input or to the development of competitively priced substitutes to the point at which dependence on new inflows of the input was ended or substantially reduced.

The speed with which the market adjusts in this way depends on several factors. The ability of technologists and businessmen to foresee a price rise might lead to the early development of substitutes to forestall the threat. On the other hand, an industry's recognition of its possible replacement by substitute products might lead it to over-produce – a policy which would result in falling profits in the short term but which might also discourage recycling and substitute development until real shortage became obvious and the industry was able to fix a price without fear of early competition.

Ownership of the scarce input and substitute materials might also become concentrated in one pair of hands or vested in one multinational corporation. Aluminium for instance, can substitute for copper in many uses and is more plentiful. Thus, a situation can be imagined where a company controlling large copper and bauxite ore resources purposely withheld the latter to gain the maximum price advantage. Concerted policy of this sort is most likely to occur where cartels operate or where one producer holds a leading or monopolistic share of the market. Using data covering twenty-one years from the end of the Second World War, Lovering (1969) has shown that consumption of mercury in the USA is not related to price but that the few producers who share the market seem to have intentionally suppressed production to encourage the government to guarantee higher minimum prices. This policy seems to have paid off as United States consumption of the metal has doubled over the period while its price has increased by more than five-fold.

Business decisions, however, are not always based purely on considerations of maximum profit available. Personalities and the status of the corporation are among the factors which complicate the issue and preclude any arbitrary interpretation of economic motives. Where a firm which controls a scarce resource not threatened by substitutes is concerned, there is no guarantee that it will suppress production to force prices up at a faster rate.

Those who make policy may be more interested in maximising production rather than maximising profit or they may be personally susceptible to the public criticism liable to be levelled at them for causing inflation.

Where property rights are not established, a situation in which production is suppressed to raise prices is likely to be reversed, notwithstanding the threat posed by substitutes or recycling. Consider the case of the fishing industry. Although environmentalists may point out the depletion of fish as a resource and British fishermen the shortsightedness of their Continental counterparts in using nets of smaller mesh size which catch the younger stock before they can breed to replacement level, likely developments are a wider adoption of this net and increased overfishing unless an equitable settlement of territorial water rights can be reached. Oilfields may pose the same sort of problem. One company will not suppress production if a rival at a neighbouring well is literally sucking the resource out from under its feet: rather it will sell at the going price irrespective of expectations of future shortage.

The market will not, then, necessarily respond in as simplistic a way as is sometimes suggested. Decisions are not taken passively in the wake of the changing pattern of demand and supply: rather these changes are anticipated and action is taken to hasten, delay or alter them. Neither is the pattern of response predictable. For both these reasons technology cannot be expected to develop in a logical stepwise way.

## 11.5 Some conclusions

What then is the outcome of the debate which has been formulated in this chapter within a polarised set of arguments, since much of the discussion in the past decade has been conducted in this way? Perhaps it is important to begin by stressing that the prognostications with respect to the availability of minerals to the end of this century evident in Chapters 5, 6 and 7 have been founded upon an examination and more often than not use of resource, reserve and projected demand figures put forward by the USBM. In the past these have been criticised as being biased in the direction of addressing US interests to the extent that once the problem of securing its own needs is somehow answered, it feels no longer compelled to pursue an exhaustive investigation of world resources and thus the forecasts have a wide margin of error (International Bank for Reconstruction and Development, 1972). Whilst such a statement may have been true in the early 1970s and to some extent distorted the basic input to the Meadows model (1972), the analytical framework being deployed by the early 1980s was of a different order. Allied to sophisticated computer data handling were the advantages for the forecaster to be gained from the utilisation of remote sensing techniques and the application of computerised modelling systems in the analysis of specific mineralised sites. These new inputs greatly improved the reliability of reserve and resource estimates. However, whether the Meadows (1972) data is under consideration or that used by the author here for 1980, whether the USBM analysis is checked against that of other experts working in the field such as those of the *Mining Annual Review* (as in this book), the extent to which world availability of a given mineral can be quantified will still to some extent remain an open question until *all* sites are known and worked out, for even intensive drilling will not reveal everything. Moreover as has been frequently reiterated in this chapter and earlier, what constitutes a resource or a reserve is not fixed over time; indeed there are even instances of so called 'waste' being viewed eventually not only as a resource but part of the reserve of a mineral.

On the demand side, forecast levels are equally questionable. It is an exercise made fraught not only by the need to estimate the likely gross national product of individual nations, both developed and less developed, but by the need to embrace a considerable range of other variables, not least changing societal needs, the possibility of technological innovation, and substitution.

But given all these difficulties of both minerals assessment and demand forecasting there remains the fundamental fact that such calculations are *only* entirely relevant for that moment in time for which they are made. To forecast industrial disaster, from the extrapolation of a data set which is not a final expression of what remains to be consumed, takes no account of the way in which society may respond through the price mechanism to the rate of consumption, the search for fresh sources, or to fulfilling its needs in some other way.

All that is being said then in the relevant chapters on the individual minerals is that by using the data for 1980 with all its limitations it is possible to suggest in the case of the metallic ores, for example, that there *may* be difficulties in meeting the demand for lead, zinc, gold, silver and uranium. The dangers here of treating reserves as more than an expression of current knowledge and society's response to that knowledge is illustrated simply by the changes that have taken place in the past in reserve estimates concerning certain key metallics. As early as 1929 a study quoted by the International Bank for Reconstruction and Development (1972) concluded that in the case of lead 'the world's resources cannot meet present demand'. But at least it qualified its remarks by saying that such a statement was dependent on the then known technology of extraction and price for the metal. However forty-three years later the Meadows *et al* study indicated a further twenty-one years of reserves *even with exponential growth*. Over this period prices had risen four-fold. However, on the gross national product assumptions of the early 1980s, the rapid decline in the use of lead in petrol (unforeseen in the early 1970s), the possibility of increased recycling and the greater use of alternative forms of batteries to those constructed from lead, gave rise to the view ten years later that only by the end of this century are supplies likely to become problematic.

Although copper is not considered likely to be in short supply by the end of the century, it is again a good example of the relationship between reserves and price. Between 1935 and 1972 reserves rose by a factor of 3.5 to 345 million tonnes. By the latter date it was reckoned by the International Bank for Construction and Development that a trebling in price would raise reserves a further twenty-five times to a level eight times greater than that of 1935. However, a combination of further price rises and the transfer of resources to the status of reserves, and a reassessment of already known reserves, had raised the reserve total by 1980 to 500 million tonnes in spite of copper consumption in the intervening years. Moreover if the mining of sea-bed nodules became a reality, as was noted in Chapter 6, a further 690 million tonnes could become available according to the assessment of potential which took place in 1980. Finally, in the case of tin the same 1929 survey that looked at lead concluded that the known reserves of tin under the price range which existed over the five years prior to that date could not satisfy the increasing demand of the industrialised nations for more than a further ten years. Forty-three years later, according to Meadows *et al*, reserves were adequate for another fifteen years *under conditions of exponential market growth*. By the early 1980s, however with likely increases in demand running as low as 1 per cent per annum, partly because of a much more economical approach to the coating of steel cans, a ten million tonne reserve out of a resource base of 37 million tonnes appeared perfectly adequate until the end of the century.

Summing up, the 1980 figures, based as they are on the improved techniques used to evaluate both reserves and resources, make it possible to suggest that across the whole range of minerals considered in this book, the *likely* shortages to be encountered before the end of the century are limited largely to a few metallics. The Meadows *et al* study probably serves only to act as a reminder of the finite nature of stock resources and of the increasing need to manage these with greater care. This involves not only an awareness of the propensity of the extraction of mineral resources, especially at leaner grades, to create pollution and of the need to reject the notion of the environment as a freely available 'sink' into which pollutants may be discharged, but the avoidance of the unnecessary sterilisation of such resources, the recycling of minerals wherever practicable and the development of methods of meeting material needs which are not energy-intensive.

Over and above these considerations that will have longer term significance, management will entail the need to develop new sources of minerals outside the developed countries which have been the primary areas of investment from the mid 1960s through the 1970s. Indeed, a substantial portion of known reserves as of 1980 (probably around 55%) is situated in the less developed countries. Thus the most important medium terms question is the extent to which the needs of the developed countries will be met from such sources given the requirement of the less developed countries to have access to considerable capital as well as skills and technology. As Chapter 4 has intimated and Mikesell (1979) has emphasised, the world's need for new capacity to mine and process minerals is becoming considerable and the finance required to support it of staggering dimensions. In his study Mikesell reckoned that investment required to create new capacity might have to rise from \$2 billion per year (the average in the second half of the 1970s), to around \$12.5 billion per year from 1980 to 2000 (1977 US dollars), with a sizeable increase needed in new investment in the last half of the period. In the less developed countries alone this could amount to \$5.5 billion a year on average. As Mikesell had estimated that external financial provision to these countries had averaged less than \$1 billion a year in the 1970s, it would need to be increased to around \$4 billion a year over the last twenty years of this century, although the major part of this average level would need to be forthcoming in the last ten years.

However, as was suggested in Chapter 4, a major barrier to the expansion of minerals production will inevitably be the incapacity of less developed countries to mobilise such considerable sums from external sources in a situation where the investment climate is not favourable. Host government controls, high levels of taxation and the threat of changes in the nature of investment agreements, may pose a threat as important as outright expropriation. In such a situation, and at a time when trading patterns of mineral development have been overtaken by a shift in the ownership of the minerals industry from international mining enterprise to host governments, new forms of agreement of the Lomé II kind will be essential. Their achievement will be the challenge of the closing decades of the twentieth century.

# Bibliography

Adams, R. G. (1980) *Metal acquisition strategy: the case for an activist consumer*. National and International Management of Mineral Resources. Joint IMM AIME meeting, May.

Anders, G., Gramm, W. P., Maurice, S. C., Smitherson, C. W. (1978) *Investment effects on the mineral industry of tax and environmental policy changes: a simulation model*. Mineral Policy Background Paper No 5, Minerals Resources Branch, Ontario Ministry of Natural Resources, Toronto, July.

*Aluminium Statistical Review*, The Aluminium Association, Economic Affairs Department, Washington, D.C.

Baldwin, R. E. (1966) *Economic Development and Export Growth: a study of Rhodesia 1920–1960*, University of California Press, Berkeley (Publications of the Bureau of Business and Economic Research, UCLA).

Banks, F. E. (1977) Natural Resource Availability – some economic aspects, *Resources Policy*, **3**, No. 1 (March), 2–12.

Barber, C. F. (1980) *Mineral Investment in an Anxious World*. Institute of Mining and Metallurgy conference, National and International Management of Mineral Resources, London.

Barnett, H. J., Morse, C. (1963) *Scarcity and Growth Resources for the Future*, Johns Hopkins Press, Baltimore.

Beamish, R. J., Van Loon, J. C., Macfarland, G. R., Lichwa, J. (1975) *The effects of Sudbury smelter emissions on lakes within the White Fish Lake Reserve*, 9th Annual Conference on Trace Substances in Environmental Health, Toronto.

Beaver, S. H. (1968) *The Geology of Sand and Gravel*, Sand and Gravel Association, London.

Beckerman, W. (1972) *Attack on the gloom doom*. Inaugural address, University College, London.

Blake, C., McDowell, D. (1967) A Local Input-Output Table, *Scottish Journal of Political Economy*, 14, 227–242.

Blakemore, H (1971) Chile, in Blakemore, H. and Smith C. T. (eds) *Latin America Geographical Perspectives*, Methuen, London.

Blunden, J. R. (1972) The decision-making process, in *Economic Geography – Industrial Location Theory*, Open University Press.

279

Bibliography

Blunden, J. R., Down, C. G., Stocks, J. (1973) *Report on the environmental problems of surface minerals extraction in North America* (unpublished). Mining Environmental Research Unit, Imperial College, London, June.

Blunden, J. R., Down, Prior, R. N., Stocks, J. (1974) *Interim Report on Surface Minerals Extraction*, Mining Environmental Research Unit, Imperial College, London.

Blunden, J. R. (1975) *The Mineral Resources of Britain: A Study in Exploitation and Planning*, Hutchinson, London.

Blunden, J. R. (1977) Resources exploitation, in *Fundamentals of Human Geography*, Open University Press.

Blunden, J. R., *et al.* (1978) *Fundamentals of Human Geography: A Reader*, Harper and Row, London.

Blunden, J. R. (1980) The positive utilisation of waste materials from mines and quarries, *Proceedings of Symposium on Surface Mining Hydrology, Sedimentology and Reclamation*, University of Kentucky, Lexington, Kentucky, Dec. 1–5.

Blunden, J. R. (1981) Urban Deprivation and Pollution, *A Guided Project Course in Human Geography*, Open University Press.

Blunden, J. R. (1982) *The Economic Impact of the Proposed Hemerdon Tungsten Mine on the Plymouth Sub-region*, Evidence to the Public Inquiry, Sept.–Oct, Plymouth.

Bosson, R., Varon, B. (1977) *The Mining Industry and the Developing Countries*, Oxford University Press.

Boulding, K. E. (1971) Environment and economics, in *Environment: Resources, Pollution and Society*, W. W. Murdoch (ed.), Sinauer Associates, Stamford, Conn., pp. 359–67.

Bragg, K. (1975) Tailings pond and lagoon treatment systems, in *Technology Transfer Seminar on Mining Effluent Regulations. Guidelines and Effluent Treatment Technology as Applied to the Base Metal, Iron Ore and Uranium Mining and Milling Industry*, Environment Canada, Ottawa.

Brown, A. J. (1972) *The Framework of Regional Economics in the United Kingdom*, Cambridge University Press, Cambridge.

Brown, L. R. (1978) *The Twenty-Ninth Day: Accommodating Human Needs and Numbers to the Earth's Resources*, W. W. Norton, New York.

Brownrigg, M. (1972) The economic impact of a new university, *Scottish Journal of Political Economy*, vol. XX (2), 123–39.

Bryant, C. R. McLellan, A. G. (1974) *A Study for the Ontario Division of Mines on the Aggregate Resources of Waterloo-South Wellington Counties: Towards Effective Planning for the Aggregate Industry*, Ontario Division of Mines, Toronto.

Butlin, J. A. (1977) Economics and Recycling, *Resources Policy*, **3**. no. 2 (June), 87–95.

Carman, J. (1977) United Nations mineral exploration activitics, 1960–1976, *Natural Resources Forum, I.*

Catlow, J. Thirwell, C. G. (1976) *Environmental Impact Analysis: A study prepared for the Secretaries of State for the Environment, Scotland and Wales*, Department of the Environment (Research Report No 11).

Chapman, K. (1976) *North Sea Oil and Gas: A Geographical Perspective*, David and Charles, Newton Abbot.

Chapman, P. F. (1974) The energy costs of producing copper and aluminium from primary sources, *Metals and Materials*, **8**, no. 2, 107–11.

280

Commission on Mining and the Environment (1972) *Report of the Commission on Mining and the Environment*, The Commission, London, (Zuckerman Report).

Conn, W. D. (1977) Waste reduction – issues and policies, *Resources Policy*, **3**, no. 1 (Mar.) 23–38.

Conroy, N., Hawley, K., Keller, W., LaFrance, C. (1975) *Influence of the atmosphere on lakes in the Sudbury area*, Ministry of the Environment (unpublished paper), Toronto.

Corner, D. C., Stafford, D. C. (1971) *China Clay Sand*, Devon County Council, Exeter.

Craig, G. R. (1974) *Toxicity studies relating to acid conditions in Lake Systems*, Proc. 21st Ontario Industrial Waste Conference, June.

Cranstone, D. A. Martin, H. L. (1973) Are ore discovery costs increasing?, paper on Canadian Mineral Exploration, Resources and Outlook, *Mineral Bulletin MR137*, Department of Energy, Mines and Resources, Ottawa, Canada.

Cranstone, D. A. (1980) Canadian Ore Discoveries 1946–78: A Continuing Record of Success, *Mineral Policy Sector Internal Report MR1 80/5*, Department of Energy, Mines and Resources, Ottawa, Canada, January.

Derry, D. R., Booth, J. K. B. (1978) Mineral discoveries and exploration expenditure – a revised review 1966–76, *Mining Magazine*, **138**, no. 5 (May), 430–3.

DeVilliers, J. (1959) *The Mineral Resources of South Africa*, Government Printer. Pretoria.

De Young, J. H. (1977) Effect of tax laws on mineral exploration in Canada, *Resources Policy*, **3**, no. 2 (June), 96–107.

Dubnie, A. (1972) *Surface Mining Practices in Canada*, Report No IC292, Department of Energy, Mines and Resources, Ottawa.

Dunham, K. C. (1974) Non-renewable mineral resources, *Resources Policy*, **1**, no. 1 (Sept), 3–13.

Dunham, K. C. (1978) World supply of non fuel minerals – the geological constraints, *Resources Policy*, **4**, no. 2 (June), 92–9.

Emerson, D. W. (1977) Australian exploration and development – comments and costs, *Bulletin of the Australian Society for Exploratory Geophysics*, **8**, no. 4 (Dec.).

Enzer, H. (1980) *Economic Assessment of Ocean Mining*, Institute of Mining and Metallurgy Conference, National and International Management of Mineral Resources, May.

European Group of Mining Companies (1976) *Raw Materials and Political Risk*, Submission to the President of the Commission of European Community.

Fischman, L., Landsberg, H. (1972) Adequacy of non fuel minerals and forest reserves, *US Commission on Population Growth and the American Future Research Reports*, vol. 111, Washington DC.

Flawn, P. T. (1966) *Mineral Resources*, Rand McNally, Chicago.

Forrester, J. W. (1976) Business structure, economic cycles, and national policy, *Futures* **8**, no. 3, (June), 195–214.

Friedmann, J. R. P. (1973) *Urbanisation, Planning and National Development*, Sage, Beverly Hills.

Garner, J. F. (ed.) (1975) *Planning Law in Western Europe*, North Holland Publishing Co, Amsterdam.

Garnick, D. H. (1970) Differential regional multiplier models, *Journal of Regional Science*, **10** (1), 35–47.

Bibliography

Gluschke, W., Shaw, J., Varon, B. (1979) *Copper, the next fifteen years* D. Reidel for the United Nations, Dordrecht, Boston, London.

Goeller, H. E., Weinberg, A. M. (1976) The age of substitutability, *Science*, **191** (4227) (20 Feb.), 683–9.

Green, D. H. (1974) *Information, perception and decision-making in the industrial location decision*, unpublished PhD thesis, University of Reading.

Grieg, M. A. (1971) The regional income and employment multiplier effects of a pulp mill and paper mill, *Scottish Journal of Political Economy*, **18** (1) (Feb.), 31–48.

Grieg, M. A. (1972) *The Economic Impact of the H.I.D.B. Investment in Fisheries*, Highland and Island Development Board.

Gutt, W., Nixon, P. J. (1980) Use of waste materials in the construction industry, *Resources Policy*, **6**, no. 1 (Mar.), 71–3.

Hall, P. (1973) England *circa* 1900, in Darby H. C. (ed.) *A New Historical Geography of England*, Cambridge University Press, pp. 674–746.

Hankinson, A. (1978) *The Investment Behaviour of the Smaller Business Unit in the Plymouth Area*, unpublished thesis, University of Bath.

Hardin, G. (1978) The tragedy of the Commons, in Blunden., J. *et al. Fundamentals of Human Geography: A Reader*, Harper and Row, London, pp. 76–83.

Hawley, J. R. (1972) *The Use, Characteristics and Toxicity of Mine – Mill Reagents in the Province of Ontario*, Ontario Ministry of the Environment, Industrial Wastes Branch, Toronto.

Henstock, M. E. (1975) Materials recovery and recycling in the USA, *Resources Policy*, **1**, no. 3 (Mar.), 171–5.

House, J. W. *et al.* (1968) *Mobility of the Northern Business Manager, Report to the Ministry of Labour*. Papers on Migration and Mobility in Northern England, no. 8, Department of Geography, University of Newcastle upon Tyne, Newcastle Upon Tyne.

Hutcheson, A. M., Hogg, A. (1974) *Scotland and Oil*, Oliver and Boyd, Edinburgh.

Hutchinson, T. C. (1973) Comparative studies of the toxicity of heavy metals to phytoplanktons and their synergistic interactions, *Water Pollution Research in Canada*, no. 8.

Hutchinson, T. C., Fedorenko, A., Fitchko, J., Kuja, A., Van Loon, J., Lichwa, J. (1975) *Movement and Compartmentation of Nickel and Copper in the Aquatic Ecosystem*, 9th Annual Conference on Trace Substances in Environmental Health, Toronto.

International Bank for reconstruction and Development (1972) *Report on Limits of Growth*, New York, (Sept.).

International Energy Agency (1982) *World Energy Outlook*, OECD, Paris.

Jones, C. F., Darkenwald, G. G. (1965) *Economic Geography*, Macmillan, New York.

Keeble, D. E. (1976) *Industrial-Location and Planning in the United Kingdom*, Methuen, London.

Lasky, S. G. (1950a) How tonnage-grade relationships help predict ore reserves, *Engineering Mining Journal*, **151**, no. 4.

Lasky, S. G. (1950b) Mineral resources appraisal by the US geological survey, *Colorado School of Mines Quarterly*, **45**, no. 1A.

Lasky, S. G. (1955) Mineral Industry futures can be predicted, *Engineering Mining Journal*, **155**, no. 9.

Lewis, T. M. McNicholl, I. H. (1978) *North Sea Oil and Scotland's Economic Prospects*, Croom Helm, London.

Linzon, S. N. (1972) Effects of sulphur oxides on vegetation, *Forestry Chronicle*, **48** (4), 182.

Llewellyn-Davies (1975) *A structure plan for Shetland*, unpublished consultant's report.

Local Government Operation Research Unit with the West Riding County Planning Dept (1973) *Where to put solid wastes – A preliminary study into the economics of coordinating*.

Lovejoy, D. (ed.) (1973) *Land Use and Landscape Planning*, Leonard Hill, Aylesbury.

Lovering, T. S. (1969) *Mineral Resources From the Land*, Resources and Man, Committee on Resources and Man, National Academy of Sciences/National Research, San Francisco, W. H. Freeman.

McDivitt, J. F., Manners, G. (1974) *Minerals and Men: and exploration of the world of minerals and metals, including some of the major problems that are posed*, John Hopkins, Baltimore.

Mackay, D. I., Mackay, G. A. (1975) *The Political Economy of North Sea Oil*, Robertson, London.

Mackenzie, B. W., Bilodeau, M. L. (1979) *Effects of Taxation of Base Metal Mining in Canada*, Centre for Resource Studies, Queen's University, Kingston, Ontario.

Mackenzie, B. W. (1980) *Effects of Sulphur Dioxide Control Costs on the Canadian Base Metal Mining Sector: a preliminary analysis*, Centre for Resource Studies, Queen's University, Kingston, Ontario.

McLellan, A. G. (1967) *The Distribution of Sand and Gravel Deposits in West Central Scotland and Some Problems Concerning their Utilization*, University of Glasgow.

McLellan, A. G. (1970) Sand and Gravel in Scotland, *Cement, Lime and Gravel*, **45** (6), June, 154–6.

McLellan, A. G., Bryant, C. R. (1975) The methodology of inventory – a practical technique for assessing provincial aggregate resources, *Canadian Institute of Mining Bulletin*. 68, no. 762.

McLellan, A. G., Yundt, S. E., Dorfman, M. L. (1979) *Abandoned pits and Quarries in Ontario*, Ontario Geological Survey, Miscellaneous Paper 79 . Ontario Ministry of Natural Resources, Toronto.

McNicoll, I. H., Walker, G. (1978) *The Shetland Economy 1976–77: Structure and Performance*, Shetland Times.

McNicholl, I. H. (1980) The Impact of Oil on the Shetland Economy, *Managerial and Decision Economics*, vol 1, no. 2.

Maddox, J. (1972) *Doomsday Syndrome: An Assault on Pessimism*, Macmillan, London.

Manners, G. (1978) The future markets for minerals: some causes of uncertainty, *Resources Policy*, **4**, no. 2 (June), 100–5.

Manpower, Services Commission, Office for Scotland (1981) *Shetland Manpower Study*, Edinburgh, August.

Meadows, D. H. *et al.* (1972) *The Limits to Growth, a report for the Club of Rome's Project on the Predicament of Mankind*, Earth Island Ltd, London.

*Metal Bulletin Handbook* Metal Bulletin plc, Worcester Park, Surrey, UK.

*Metal Statistics (1970–1980)* 68th edition, Metallgesellschaft AG, Frankfurt am Main.

Bibliography

Mikesell, R. F. (1979) *New Patterns of World Mineral Development*, British–North American Committee, London.

*Mineral Facts and Problems*, Bureau of Mines, US Department of the Interior, US Government Printing Office, Washington, DC.

Mineral Resources Branch, Division of Mines (1977) *The Ontario Metal Mining Industry, Present and Future*, Ministry of Natural Resources, Ottawa.

*Minerals Year Book* Vols 1–3, Bureau of Mines, US Department of the Interior, US Government Printing Office, Washington DC.

*Mining Journal Annual Review* (1981) Mining Journal Ltd, London.

*Mining Journal Annual Review* (1982) Mining Journal Ltd, London.

Ministry of Housing and Local Government (1960) *Control of Mineral Working*, London, HMSO.

Moses, V. (1983) Breeding bugs into the system, *Proceedings of the British Association for the Advancement Of Science Annual Meeting*, University of Sussex.

Natural Environmental Research Council (1982) *United Kingdom Mineral Statistics (1981)* Institute of Geological Sciences, HMSO, London.

Nelson, J. G. (1981) *The Scottish and Alaskan Offshore Oil and Gas Experience and the Canadian Beaufort Sea*, Canadian Arctic Resources Committee, Ottawa.

Nicholl, R. E. (1977) *Energy and the Environment*, Royal Institute of Chartered Surveyors, Annual Conference Papers, London.

Odell, P., Vallenilla, L. (1978) *The Pressures of Oil: a strategy for economic revival*, Harper and Row, London.

Odell, P. R. (1966) What will gas do to the East Coast?, *New Society*, **188**, no. 5 (5 May), 8–9.

Odell, P. R. (1973) Major themes in Latin America's economic geography, in P. R. Odell, D. A. Preston, (eds) *Economies and Societies in Latin America: a Geographical Interpretation*, London, John Wiley.

OECD Committee Report (1973) *Oil: The Present Situation and Future Prospects*, OECD, Paris.

Oil and Gas Journal (1982) Penwill Publishing Company, Tulsa, Oklahoma.

Outtrim, C. P., Evans, R. G. (1978) Alberta's Oil Sands Reserves and their Evaluation in *The Oil Sands of Canada–Venezuela*, 1977, Redford D. A. and Winestock A. G. eds, Canadian Institute of Mining, Special Volume 17, Montreal.

Papanicolaou, N. A., Mackenzie, B. W. (1981) *Effects of Water Pollution Control Costs on Base Metal Mining In Canada*, Centre for Resource Studies Working Paper No 23, Queen's University, Kingston, Ontario, April.

Perry, R. (1982) *A regional economic analysis with special reference to Cornwall*. Unpublished doctoral thesis, Open University.

Place, J. L. (1972) Man's role in geomorphic change on the shoreline of Los Angeles County, California, in *Papers Submitted to the Twenty-Second International Geographical Congress, Canada*, University of Toronto Press, Toronto. 2 vols, 655–6.

Poland, J. F., Davis, G. H. (1969) Land subsidence due to withdrawal of fluids, *Reviews in Engineering Geology*, **2**, Geological Society of America, Boulder, Colorado.

Pounce, R. Upson, R. Walker, C. (1976) *Scottish Industry and North Sea Oil*. Department of Trade and Industry.

Prast, W. G. (1979) The role of the United Nations in mineral exploitation, *Mining Magazine*, **141**, No 2, London (August), 114–7.

*Presidents Commission on Coal* (1980) Energy Information Administration, Washington DC.

*Quarterly Iron and Steel Bulletin* Statistical Office of the European Community, Brussels.

Radetzki, M., Zorn, S. (1979) *Financing Mining Projects in Developing Countries: A United Nations Study*, Mining Journal Books, London.

Rajaraman, I. (1976) Non-renewable resources: a review of long-term projections, *Futures*, **8**, No. 3 (June) 228–42.

Redford, D. A., Winestock, A. G. (1978). *Oil Sands of Canada–Venezuela*, Canadian Institute of Mining, Special Volume **17**, Montreal.

Richardson, H. W. (1976) *Regional and Urban Economics*, Pitman, London.

Ridley, P. W. (1980) '*Role of governments in relation to minerals, nationally and internationally*'. Paper given at the Institute of Mining and Metallurgy Conference on The National and International Management of Mineral Resources, London (May).

Ripley, E. A. *et al.* (1978) Environmental Impact of Mining in Canada, (*National Impact of Mining Series, No 7*) Centre for Resource Studies, Queen's University, Kingston, Ontario.

Roberts, F. (1973) Future resources for engineering materials, *Chartered Mechanical Engineer*, **19** (April).

Russell, W. (1972) *In Great Waters: report on a study of the economic and social impact of the Highlands and Islands Development Board's Investment in Fisheries*, Highlands and Islands Development Board, Inverness. 2 vols. **2**: Greig, M. A. *The economic impact of the Highlands and Islands Development Board investment in fisheries – economic (multiplier) assessment.*

Sandbach, F. (1982) *Principles of Pollution Control*, Longman, London.

Saunders, J. O. (1980) *The Law of the Sea and Canada's Mineral Industry*, Centre for Resource Studies, Queen's University, Kingston, Ontario.

Scottish Economic Bulletin (1977) *The Impact of North Sea Oil-Related Activity on Employment in Scotland*, No 11.

Scottish Information Unit (1975) *Factsheet*, March.

Shank, R. J. (1980) *Dynamics of a national resource assessment programme*. Paper given at the Institute of Mining and Metallurgy Conference on The National and International Management of Mineral Resources, London (May).

Shetland Islands Council Research and Development Department (1978) *Shetland Oil Era Phase 2*.

Shetland Islands Council Research and Development Department (1980) *Shetland in Statistics No 9*.

Singer, D. A. (1977) Long-term adequacy of metal resource, *Resources Policy*, **3**, No 2, 127–33.

Solid Waste Task Force (1974) *Report to the Ontario Ministry of the Environment*, Toronto.

Solow, R. M. (1974) The economics of resources or the resources of economics (The Richard T. Ely Memorial Lecture), *American Economic Review* **64** No 2 1–14.

Spooner, D. (1981) *Mining and Regional Development*, Oxford University Press, Oxford.

Standing Conference on London and South East Regional Planning (1976) *The Improvement of London's Green Belt: report by the Green Belt Working Party*, London.

Stevens, Sir Roger, (Chairman) (1976) *Report of the Committee on Planning Con-*

*trol Over Mineral Working*, HMSO, London.

Stokes, P. M. *et al.* (1973) Heavy-metal tolerance in algae isolated from contaminated lakes near Sudbury, Ontario, *Canadian Journal of Botany*, **51**, No 11, 2155–68.

Taylor, D. W. (ed.) (1974) Offshore oil: a cause for regret? *Architects' Journal* (Special Issue), **159**, No 26, 26 June.

Thomas, I. A. (1980) *Influence of land-use planning on availability of minerals*. Paper given at the Institute of Mining and Metallurgy Conference on The National and International Management of Mineral Resources, London, (May).

Thomas, L. J. (1973) *An Introduction to Mining, Exploration, Feasibility, Extraction and Rock Mechanics*, Halsted Press, New York.

Townroe, P. M. (1971) *Industrial location decisions: a study in management behaviour*, (Occasional Paper 15). University of Birmingham, Centre for Urban and Regional Studies, Birmingham.

Tugendhat, C. (1978) Raw materials – the third world and the European Community *Transactions of the Institute of Mining and Metallurgy* (Section A, Mining Industry 87).

United Nations Statistical Year Books.

Van Rensburg, W. C. J. *et al.* (1969) *South Africa's Coal resources*, Coal Advisory Board, Pretoria.

Van Rensburg, W. C. J. (1975) 'Reserves' as a leading indicator to future mineral production, *Resources Policy*, **1**, No 6 (December), 343–56.

Venter, F. A. (1952), 'Coal in the Union of South Africa', *Transactions. Geological Society of South Africa, LV.*

Walker, K. J. (1979) Materials consumption implications of a fully industrialised world, *Resources Policy*, **5**, No 4 (December), 242–59.

Wargo, J. G. (1973) Trends of corporate mineral exploration expenditure 1968–1971, *Mining Engineering*, New York, 25 May.

Warren, K. (1973) *Mineral Resources*, David and Charles, Newton Abbot.

Willox, W. A. (1980) *Exploration in the third world: the role of the consultant and implications for national policies*. Paper given at the Institute of Mining and Metallurgy Conference on The National and International Management of Mineral Resources, London, May.

Wixson, B. G. (1972) *The Lead Industry as a Source of Trace Metals in the Environment,* Environmental Resources Conference, Ohio.

World Bank, (1977) *Annual Report*, Washington DC.

World Coal Study (1980) *Coal-Bridge to the Future*, WOCOL, London.

Wybergh, W. J. (1928) *Coal Resources of the Union of South Africa*, Geological Survey, Memoir 19.

Zimmermann, E. W. (1951) *World Resources and Industries; a functional appraisal of the availability of agricultural and industrial materials*, Harper and Row, New York, revised edition.

Zuckerman, Lord (1972) Science, technology and environmental management, *The Environment This Month*, August.

Zwartendyk, J. (1972) What is mineral endowment and how should we measure it? *Canadian Mineral Resources Branch, Mineral Bulletin*, **126**.

# Index

# Index

Index

# Index

zinc, 142–3
non-metalliferous
barytes, 104–5
cement, 75
fluorspar, 98, 101–3
gypsum, 80
kaolin/ball clay, 109
phosphate, 88
salt, 96–7
sulphur, 92–4
oil, 176–8, 182
Mikesell, R. F., 277–8
models, theoretical and employment
multiplier, 231–4
molybdenum, 47, 51, 58, 61, **120, 122,
124–5, 127**, 263
Mongolia
coal, 186
non-metalliferous
fluorspar, 100–1, 103
rare earths, 160
monitoring reserves, 22–3
Morocco
non-ferrous metallics
antimony, 158
cobalt, 123
lead, 142
non-metalliferous
barytes, 104–5
phosphate, 85–6, 88–9
salt, 60
Morse, C., 272
Moses, V., 270
multinational companies, 52
multiplier model, 231–4
case studies
oil, 242–52
tungsten, 234–42
regional estimates, 252–5


Namibia
non-ferrous metallics
lithium, 161
uranium, 170–2
nationalism and investment, 52–3
natural gas, 40, **180–3**, 204
see also oil
Nelson, J. E., 248
Netherlands, 62, 170, 208, 213
iron, 48, 116, 118
non-ferrous metallics
aluminum, 130
cadmium, 163
zinc, 141
non-metalliferous
cement, 75
salt, 96–7
sulphur, 93
oil and natural gas, 180–2

296

New Zealand, 43, 213
iron, 114
non-ferrous metallic aluminium, 130
Nicholl, R. E., 43–4
nickel, **120, 122, 126–7**
environmental problems, 195–7, 201
reserves and resources, 33, 47, 61,
262–3, 265, 272
supply and demand, 51–2, 58
Niger, 61
non-ferrous metallic uranium, 170–2
Nigeria, 61
non-ferrous metallic tin, 140, 144
oil, 175–6, 182
Nixon, P. J., 215–16
noise pollution, 70, 191
non-ferrous metallics, 36, 45–6, 50–54, 56,
120–73
*see also* aluminium, antimony, cadmium,
chromite, cobalt, copper, gold,
indium, lead, lithium, magnesium,
manganese, mercury,
molybdenum, nickel, platinum,
rare earths, rhenium, selenium,
silver, tellurium, tin, titanium,
tungsten, uranium, vanadium, zinc
non-metalliferous minerals
localised, 45–6, **78–112**
*see also* asbestos, barytes, boron,
fluorspar, gypsum, kaolin/ball
clay, phosphate, potash, salt,
sulphur
ubiquitous, 44–6, 49, 56, **63–112**
*see also* cement, clay, common, gravel,
sand, shales, stone
North Sea oil, 46, 242–54, 270
*see also* oil
Norway, 206
iron, 114
non-ferrous metallics, 129–30, 135–6,
138, 142
non-metalliferous sulphur, 93
oil, 176
nuclear power *see* uranium

ocean mining, 264–5
non-ferrous metallics, 127, 154
non-metalliferous, 68–9
Odell, P. R., 40–42, 229, 252
oil, **174–80, 183**
environmental problems, 180, 204
price rises, 56, 65
regional development, 229, 242–52
reserves and resources, 20, 33, 39–41,
46–7, 261–2, 269, 267–71
supply and demand, 53, 56, 177–9
supply, future, 267–71
'synthetic', 267
Oman, oil, 176, 182

# Index

resources and reserves
classification, 29–49
categorization of, 29–36
stock
availability, 36–45
classifying, 45–9
defining, 18–28
case studies, 24–8
human appraisal, 18–21
theoretical considerations, 21–4
*see also* future availability, *and
individual minerals and countries*
rhenium, **166–7**
Richardson, H. W., 256
Ridley, P. W., 59
Ripley, E. A., 196
Ripley, M., 4, 8, 17
Roberts, F., 266
rock *see* stone
Romania
iron, 114, 118
non-ferrous metallics
aluminium, 130
lead, 142
zinc, 142
non-metalliferous
cement, 75
kaolin/ball clay, 109
salt, 97
sulphur, 93

saleable product, waste as, 211–16
salt, 33, 46, 60, **94–8**, 203
sand, 31, 44–5, 65, **66–9**, **71**, 191, 208–9,
211
Sandbach, F., 83, 196
Sardinia, non-metalliferous fluorspar, 98
Saudi Arabia
non-metalliferous sulphur, 93
oil, 175–6
Saunders, J. O., 265
seabed mining *see* ocean
selenium, 33, 58, **166–8**
Senegal, 61
non-metalliferous phosphate, 86
Seychelles, 61
shales, **76–8**, 215
Shank, R. J., 21–6
Shaw, J., 59
Sierra Leone, 61
non-ferrous metallics
aluminium, 130
titanium, 137–8
silver, 34, 47, 51, 61, **150–4**
Singer, D. A., 273
smelting, 196–202
*see also* extraction and processing
Smith, E., 195
soil pollution, 210

solar energy, 42, 125, 271
Solow, R. M., 266
Somalia, 61
South Africa
coal, 28, 187–8, 213
investment, 56, 58–9
iron, 114–15
non-ferrous metallics
aluminium, 130
antimony, 157–8
chromite, 121, 124, 126
copper, 140
gold, 46, 149–50, 213, 255
lead, 142
magnesium, 135
manganese, 121, 126
nickel, 122
platinum, 150, 154–6
tin, 140
titanium, 136–8
uranium 169–70, 172
vanadium, 123
zinc, 142
non-metalliferous
asbestos, 80–2
cement, 75
fluorspar, 98, 100–1, 103
kaolin/ball clay, 109
phosphate, 86, 88
salt, 97
sulphur, 93
oil, 185–6
waste utilisation, 213
Soviet Union *see* Union of Soviet, *etc.*
Spain, 170
iron, 48, 114, 118
non-ferrous metallics
aluminium, 130
gold, 150
lead, 142
mercury, 163–4
silver, 150
zinc, 142
non-metalliferous
cement, 75
fluorspar, 98
kaolin/ball clay, 109
salt, 97
spatial patterns of mining investment, 58–9
spoil heaps *see* waste
Spooner, D., 46, 229, 255–6
spreading of mineral reserves, 264
Sri Lanka
non-ferrous metallics
rare earths, 160
titanium, 137
Stafford, D. C., 224
Steel *see* iron and steel
stock resources, 29–30
availability, 36–45

298

Index